New York State Learning Standards

and Success Strategies for the State Test

This book is customized for New York and the lessons match the *New York State Learning Standards*. The Measuring Up® program includes comprehensive worktexts and New York Diagnostic Practice Tests, which are available separately.

800-822-1080
www.PeoplesEducation.com

Executive Vice President, Chief Creative Officer: Diane Miller
Editorial Development: Navta Associates, Inc.
Managing Editor: Kerri Gerro
Vice President of Marketing: Victoria Ameer Kiely
Senior Marketing Manager: Christine Grasso
Associate Marketing Manager: Victoria Leo
Production Director: Jason Grasso
Production Manager: Jennifer Bridges
Assistant Production Managers: Jennifer Tully, Steven Genzano
Copy Editors: Shelly Rawson, Dee Josephson, Katy Leclercq
Director of Permissions: Kristine Liebman
Illustrators: Armando Báez, Sal Esposito, Bob J. Eckhardt, Sharon MacGregor, Dan Lish, Karen's Custom Airbrush

New York Advisory Panel

Steve Steuerman, Science Curriculum Specialist—Community School District #22, New York, NY

Rose Villani, Director of Science—District #11, New York, NY

Jim Plansky, Assistant Principal at Walt Whitman—Huntington Station, NY

Ron Schick, Former Coordinator of Science—Yonkers, NY

Judith Fix, Science Mentor—Buffalo, NY

Robert Hansen, Teacher, Zena Elementary/Kingston City Schools—Kingston, NY

Peoples Education, Inc.
299 Market Street
Saddle Brook, New Jersey 07663

ISBN 978-1-4138-7465-5

Printed in the United States of America.

10 9 8 7 6 5 4 3 2 1

Note: Standards in the Living Environment section of the Science Standard 4 are indicated with the letters **LE** preceding the Major Understandings. Standards from the Physical Setting section are indicated with the letter **PS** preceding the Major Understandings.

CHAPTER 1 Scientific Inquiry (continued)

Build your intellectual stamina with these activities. Each question will apply multiple standards on a higher level. This section gives you a chance to show what you know about the processes needed for science investigations, and strengthen your test-taking abilities.

CHAPTER 2 Living Things

Build your intellectual stamina with these activities. Each question will apply multiple standards on a higher level. This section gives you a chance to show what you know about living things, and strengthen your test-taking abilities.

Note: Standards in the Living Environment section of the Science Standard 4 are indicated with the letters LE preceding the Major Understandings. Standards from the Physical Setting section are indicated with the letter PS preceding the Major Understandings.

CHAPTER 3 Growth and Development

Build your intellectual stamina with these activities. Each question will apply multiple standards on a higher level. This section gives you a chance to show what you know about life cycles and growth of living things, and strengthen your test-taking abilities.

CHAPTER 4 Ecology

Build your intellectual stamina with these activities. Each question will apply multiple standards on a higher level. This section gives you a chance to show what you know about ecology and environmental science, and strengthen your test-taking abilities.

Note: Standards in the Living Environment section of the Science Standard 4 are indicated with the letters LE preceding the Major Understandings. Standards from the Physical Setting section are indicated with the letter PS preceding the Major Understandings.

CHAPTER 5 Earth

Build your intellectual stamina with these activities. Each question will apply multiple standards on a higher level. This section gives you a chance to show what you know about the changes that occur on Earth, and strengthen your test-taking abilities.

CHAPTER 6 Matter

Build your intellectual stamina with these activities. Each question will apply multiple standards on a higher level. This section gives you a chance to show what you know about the properties of matter, and strengthen your test-taking abilities.

Note: Standards in the Living Environment section of the Science Standard 4 are indicated with the letters **LE** preceding the Major Understandings. Standards from the Physical Setting section are indicated with the letter **PS** preceding the Major Understandings.

CHAPTER 7 Energy

CHAPTER 8 Force and Motion

Note: Standards in the Living Environment section of the Science Standard 4 are indicated with the letters **LE** preceding the Major Understandings. Standards from the Physical Setting section are indicated with the letter **PS** preceding the Major Understandings.

Correlation to the New York State Learning Standards and Major Understandings

This worktext is customized to the *New York Elementary Science Core Curriculum* and will help you prepare for the *Grade 4 Elementary-Level Science Test.*

After the lesson is completed, place a (✓) to indicate Mastery or a (**X**) to indicate Review Needed.

Chapter 1: Scientific Inquiry									
	Review Skill								
	Mastered Skill								
	Lessons	1	2	3	4	5	NA	Chap Review	End Rev
M1.1	**Mathematical Analysis** Use special mathematical notation and symbolism to communicate in mathematics and to compare and descibe quantities, express relationships, and relate mathematics to their immediate environment.								
M1.1a	Use plus, minus, greater than, less than, equal to, multiplication, and division signs.	❍	❍	❍	❍	★		★	★
M1.1b	Select the appropriate operation to solve mathematical problems.	❍	❍	❍	❍	★		★	✓
M1.1c	Apply mathematical skills to describe the natural world.	❍	❍	❍	❍	★		★	✓
M2.1	**Mathematical Analysis** Use simple logical reasoning to develop conclusions, recognizing that patterns and relationships present in the environment assist them in reaching their conclusions.								
M2.1a	Explain verbally, graphically, or in writing the reasoning used to develop mathematical conclusions.	❍	❍	❍	❍	★		★	✓
M2.1b	Explain verbally, graphically, or in writing patterns and relationships observed in the physical and living environment.	❍	❍	❍	❍	★		★	✓
M3.1	**Mathematical Analysis** Explore and solve problems generated from school, home, and community situations, using concrete objects or manipulative materials when possible.								
M3.1a	Use appropriate scientific tools, such as metric rulers, spring scale, pan balance, graph paper, thermometers [Fahrenheit and Celsius], graduated cylinder to solve problems about the natural world.	❍	★	✓	✓	✓		★	✓
S1.1	**Scientific Inquiry** Ask "why" questions in attempts to seek greater understanding concerning objects and events they have observed and heard about.								
S1.1a	Observe and discuss objects and events and record observations.	★	✓	✓	✓	✓		★	✓
S1.1b	Articulate appropriate questions based on observations.	★	✓	✓	✓	✓		★	✓
S1.2	**Scientific Inquiry** Question the explanations they hear from others and read about, seeking clarification and comparing them with their own observations and understandings.								
S1.2a	Identify similarities and differences between explanations received from others or in print and personal observations or understandings.	★	✓	✓	✓	✓		★	✓
S1.3	**Scientific Inquiry** Develop relationships among observations to construct descriptions of objects and events and to form their own tentative explanations of what they have observed.								
S1.3a	Clearly express a tentative explanation or description which can be tested.	★	✓	✓	✓	✓		✓	✓

Note: Standards in the Living Environment section of the Science Standard 4 are indicated with the letters LE preceding the Major Understandings. Standards from the Physical Setting section are indicated with the letter PS preceding the Major Understandings.

★ **Standard Covered** ❍ **Standard to be covered** ✓ **Standard previously covered**

Correlation to the New York State Learning Standards and Major Understandings

This worktext is customized to the *New York Elementary Science Core Curriculum* and will help you prepare for the *Grade 4 Elementary-Level Science Test.*

After the lesson is completed, place a (✓) to indicate Mastery or a (**X**) to indicate Review Needed.

Chapter 1: Scientific Inquiry (Continued)									
	Review Skill								
	Mastered Skill								
	Lessons	1	2	3	4	5	NA	Chap Review	End Rev
S2.1	**Scientific Inquiry** Develop written plans for exploring phenomena or for evaluating explanations guided by questions or proposed explanations they have helped formulate.								
S2.1a	Indicate materials to be used and steps to follow to conduct the investigation and describe how data will be recorded (journal, dates and times, etc.).	○	★	★	✓	✓		★	✓
S2.2	**Scientific Inquiry** Share their research plans with others and revise them based on their suggestions.								
S2.2a	Explain the steps of a plan to others, actively listening to their suggestions for possible modification of the plan, seeking clarification and understanding of the suggestions and modifying the plan where appropriate.	○	★	✓	✓	✓		★	✓
S2.3	**Scientific Inquiry** Carry out their plans for exploring phenomena through direct observation and through the simple instruments that permit measurement of quantities, such as length, mass, volume, temperature, and time.								
S2.3a	Use appropriate "inquiry and process skills" to collect data.	○	★	★	✓	✓		★	✓
S2.3b	Record observations accurately and concisely.	○	★	✓	✓	✓		★	✓
S3.1	**Scientific Inquiry** Organize observations and measurements of objects and events through classification and preparation of simple charts and tables.								
S3.1a	Accurately transfer data from a science journal or notes to appropriate graphic organizer.	○	○	○	★	✓		★	✓
S3.2	**Scientific Inquiry** Interpret organized observations and measurements, recognizing simple patterns, sequences, and relationships.								
S3.2a	State, orally and in writing, any inferences or generalizations indicated by the data collected.	○	○	○	★	✓		★	✓
S3.3	**Scientific Inquiry** Share their findings with others and actively seek their interpretations and ideas.								
S3.3a	Explain their findings to others, and actively listen to suggestions for possible interpretations and ideas.	○	○	○	★	✓		★	✓
S3.4	**Scientific Inquiry** Adjust their explanations and understandings of objects and events based on their findings and new ideas.								
S3.4a	State, orally and in writing, any inferences or generalizations indicated by the data, with appropriate modifications of their original prediction/explanation.	○	○	○	★	✓		★	✓
S3.4b	State, orally and in writing, any new questions that arise from their investigation.	○	○	○	★	✓		★	✓

★ **Standard Covered** ○ **Standard to be covered** ✓ **Standard previously covered**

Correlation to the New York State Learning Standards and Major Understandings

This worktext is customized to the *New York Elementary Science Core Curriculum* and will help you prepare for the *Grade 4 Elementary-Level Science Test.*

After the lesson is completed, place a (✓) to indicate Mastery or a (**X**) to indicate Review Needed.

Chapter 2: Living Things		6	7	8	9	10	11	Chap Review	End Rev
	Review Skill								
	Mastered Skill								
	Lessons	6	7	8	9	10	11	Chap Review	End Rev
LE1.1	Describe the characteristics of and variations between living and nonliving things.								
LE1.1a	Animals need air, water, and food in order to live and thrive.	★	✓	✓	✓	✓	✓	★	✓
LE1.1b	Plants require air, water, nutrients, and light in order to live and thrive.	★	✓	✓	✓	✓	✓	✓	✓
LE1.1c	Nonliving things do not live and thrive.	★	✓	✓	✓	✓	✓	✓	✓
LE1.1d	Nonliving things can be human-created or naturally occurring.	★	✓	✓	✓	✓	✓	★	✓
LE1.2	Describe the life processes common to all living things.								
LE1.2a	Living things grow, take in nutrients, breathe, reproduce, eliminate waste, and die.	★	✓	✓	✓	✓	✓	★	✓
LE2.1	Recognize that traits of living things are both inherited and acquired or learned.								
LE2.1a	Some traits of living things have been inherited.	❍	★	✓	✓	✓	✓	★	✓
LE2.1b	Some characteristics result from an individual's interactions with the environment and cannot be inherited by the next generation.	❍	★	✓	✓	✓	✓	★	✓
LE2.2	Recognize that for humans and other living things there is a genetic continuity between generations.								
LE2.2a	Plants and animals closely resemble their parents and other individuals in their species.	❍	★	✓	✓	✓	✓	★	✓
LE2.2b	Plants and animals can transfer specific traits to their offspring when they reproduce.	❍	★	✓	✓	✓	✓	★	✓
LE3.1	Observe and describe properties of materials, using appropriate tools.								
LE3.1a	Each animal has different structures that serve different functions in growth, survival, and reproduction.	❍	❍	★	✓	✓	✓	★	✓
LE3.1b	Each plant has different structures that serve different functions in growth, survival, and reproduction.	❍	❍	❍	★	✓	✓	★	✓
LE3.1c	In order to survive in their environment, plants and animals must be adapted to that environment.	❍	❍	❍	❍	★	✓	★	✓
LE3.2	Observe that differences within a species may give individuals an advantage in surviving and reproducing.								
LE3.2a	Individuals within a species may compete with each other for food, mates, space, water, and shelter in their environment.	❍	❍	❍	❍	★	✓	★	✓
LE3.2b	All individuals have variations, and because of these variations individuals of a species may have an advantage in surviving and reproducing.	❍	❍	❍	❍	★	✓	★	✓
LE5.1	Describe basic life functions of common living specimens.								
LE5.1a	All living things grow, take in nutrients, breathe, reproduce, and eliminate waste.	★	✓	✓	✓	✓	✓	★	✓
LE5.1b	An organism's external physical features can enable it to carry out life functions in its particular environment.	❍	❍	★	★	✓	✓	★	✓
LE5.2	Describe some survival behaviors of common living specimens.								

★ **Standard Covered** ❍ **Standard to be covered** ✓ **Standard previously covered**

Correlation to the New York State Learning Standards and Major Understandings

This worktext is customized to the *New York Elementary Science Core Curriculum* and will help you prepare for the *Grade 4 Elementary-Level Science Test.*

After the lesson is completed, place a (✓) to indicate Mastery or a (**X**) to indicate Review Needed.

Chapter 2: Living Things (Continued)									
	Review Skill								
	Mastered Skill								
	Lessons	6	7	8	9	10	11	Chap Review	End Rev
LE5.2a	Plants respond to changes in their environment.	❍	❍	❍	❍	❍	★	★	✓
LE5.2b	Animals respond to change in their environment.	❍	❍	❍	❍	❍	★	★	✓
LE5.2c	Senses can provide essential information (regarding anger, food, mates, etc.) to animals about their environment.	❍	❍	❍	❍	❍	★	★	✓
LE5.2d	Some animals, including humans, move from place to place to meet their needs.	❍	❍	❍	❍	❍	★	★	✓
LE5.2e	Particular animal characteristics are influenced by changing environmental conditions including: fat storage in winter, coat thickness in winter, camouflage, shedding of fur.	❍	❍	❍	❍	❍	★	★	✓
LE5.2f	Some animal behaviors are influenced by environmental conditions. These behaviors may include: nest building, hibernating, hunting, migrating, and communicating.	❍	❍	❍	❍	❍	★	★	✓

Chapter 3: Growth and Development									
	Review Skill								
	Mastered Skill								
	Lessons	12	13	14	15	NA	NA	Chap Review	End Rev
LE4.1	Describe the major stages in the life cycles of selected plants and animals.								
LE4.1a	Plants and animals have life cycles. These may include beginning of a life, development into an adult, reproduction as an adult, and eventually death.	★	★	✓	✓			★	✓
LE4.1b	Each kind of plant goes through its own stages of growth and development that may include seed, young plant, and mature plant.	★	✓	✓	✓			★	✓
LE4.1c	The length of time from beginning of development to death of the plant is called its life span.	★	✓	✓	✓			★	✓
LE4.1d	Life cycles of some plants include changes from seed to mature plant.	★	✓	✓	✓			★	✓
LE4.1e	Each generation of animals goes through changes in form from young to adult. This completed sequence of changes in form is called a life cycle. Some insects change from egg to larva to pupa to adult.	❍	★	✓	✓			★	✓
LE4.1f	Each kind of animal goes through its own stages of growth and development during its life span.	❍	★	✓	✓			★	✓
LE4.1g	The length of time from an animal's birth to its death is called its life span. Life spans of different animals vary.	❍	★	✓	✓			★	✓
LE4.2	Describe evidence of growth, repair, and maintenance, such as nails, hair, and bone, and the healing of cuts and bruises.								
LE4.2a	Growth is the process by which plants and animals increase in size.	★	★	★	✓			★	✓
LE4.2b	Food supplies the energy and materials necessary for growth and repair.	★	★	★	✓			✓	✓
LE5.2	Describe some survival behaviors of common living specimens.								
LE5.2g	The health, growth, and development of organisms are affected by environmental conditions such as the availability of food, air, water, space, shelter, heat, and sunlight.	★	★	✓	✓			★	✓
LE5.3	Describe the factors that help promote good health and growth in humans.								

★ **Standard Covered** ❍ **Standard to be covered** ✓ **Standard previously covered**

Correlation to the New York State Learning Standards and Major Understandings

This worktext is customized to the *New York Elementary Science Core Curriculum* and will help you prepare for the *Grade 4 Elementary-Level Science Test*.

After the lesson is completed, place a (✓) to indicate Mastery or a (**X**) to indicate Review Needed.

Chapter 3: Growth and Development (Continued)									
	Review Skill								
	Mastered Skill								
	Lessons	12	13	14	15	NA	NA	Chap Review	End Rev
LE5.3a	Humans need a variety of healthy foods, exercise, and rest in order to grow and maintain good health.	❍	❍	❍	★			★	✓
LE5.3b	Good health habits include hand washing and personal cleanliness; avoiding harmful substances (including alcohol, tobacco, illicit drugs); eating a balanced diet; engaging in regular exercise.	❍	❍	❍	★			★	✓

Chapter 4: Ecology									
	Review Skill								
	Mastered Skill								
	Lessons	16	17	18	NA	NA	NA	Chap Review	End Rev
LE6.1	Describe how plants and animals, including humans, depend upon each other and the nonliving environment.								
LE6.1a	Green plants are producers because they provide the basic food supply for themselves and animals.	★	✓	✓				★	✓
LE6.1b	All animals depend on plants. Some animals (predators) eat other animals (prey).	★	✓	✓				★	✓
LE6.1c	Animals that eat plants for food may in turn become Food for other animals. This sequence is called a food chain.	★	✓	✓				★	✓
LE6.1d	Decomposers are living things that play a vital role in recycling nutrients.	★	✓	✓				★	✓
LE6.1e	An organism's pattern of behavior is related to the nature of that organism's environment, including the kinds and numbers of other organisms present, the availability of food and other resources, and the physical characteristics of the environment.	❍	★	✓				★	✓
LE6.1f	When the environment changes, some plants and animals survive and reproduce, and others die or move to new locations.	❍	★	✓				★	✓
LE6.2	Describe the relationship of the Sun as an energy source for living and nonliving cycles.								
LE6.2a	Plants manufacture food by utilizing air, water, and energy from the Sun.	★	✓	✓				★	✓
LE6.2b	The Sun's energy is transferred on Earth from plants to animals through the food chain.	★	✓	✓				★	✓
LE6.2c	Heat energy from the Sun powers the water cycle.	❍	★	✓				★	✓
LE7.1	Identify ways in which humans have changed their environment and the effects of those changes.								
LE7.1a	Humans depend on their natural and constructed environments.	❍	❍	★				★	✓
LE7.1b	Over time humans have changed their environment by cultivating crops and raising animals, creating shelter, using energy, manufacturing goods, developing means of transportation, changing populations, and carrying out other activities.	❍	❍	★				★	✓
LE7.1c	Humans, as individuals or communities, change environments in ways that can be either helpful or harmful for themselves and other organisms.	❍	❍	★				★	✓

★ **Standard Covered** ❍ **Standard to be covered** ✓ **Standard previously covered**

Correlation to the New York State Learning Standards and Major Understandings

This worktext is customized to the *New York Elementary Science Core Curriculum* and will help you prepare for the *Grade 4 Elementary-Level Science Test*.

After the lesson is completed, place a (✓) to indicate Mastery or a (**X**) to indicate Review Needed.

Chapter 5: Earth									
	Review Skill								
	Mastered Skill								
	Lessons	19	20	21	22	23	NA	Chap Review	End Rev
PS1.1	Describe patterns of daily, monthly, and seasonal changes in the atmosphere.								
PS1.1a	Natural cycles and patterns include: • Earth spinning around once every 24 hours (rotation), resulting in day and night; • Earth moving in a path around the Sun (revolution), resulting in one Earth year; • the length of daylight and darkness varying with the seasons; • weather changing from day to day and through the seasons; • the appearance of the Moon changing as it moves in a path around Earth to complete a single cycle.	❍	❍	❍	❍	★		★	✓
PS1.1b	Humans organize time into units based on natural motions of Earth: second, minute, hour, week, month.	❍	❍	❍	❍	★		★	✓
PS1.1c	The Sun and other stars appear to move in a recognizable pattern both daily and seasonally.	❍	❍	❍	❍	★		★	✓
PS2.1	Describe the relationship among air, water, and land on Earth.								
PS2.1a	Weather is the condition of the outside air at a particular moment.	★	✓	✓	✓	✓		★	✓
PS2.1b	Weather can be described and measured by: • temperature • wind speed and direction • form and amount of precipitation • general sky conditions (cloudy, sunny, partly cloudy)	★	✓	✓	✓	✓		★	✓
PS2.1c	Water is recycled by natural processes on Earth. • evaporation: changing of water (liquid) into water vapor (gas) • condensation: changing of water vapor (gas) into water (liquid) • precipitation: rain, sleet, snow, hail • runoff: water flowing on Earth's surface • groundwater: water that moves downward into the ground	❍	★	✓	✓	✓		★	✓
PS2.1d	Erosion and deposition result from the interaction among air, water, and land. • interaction between air and water breaks down earth materials • pieces of earth material may be moved by air, water, wind, and gravity • pieces of earth material will settle or deposit on land or in the water in different places • soil is composed of broken-down pieces of living and nonliving earth material	❍	❍	★	✓	✓		★	✓
PS2.1e	Extreme natural events (floods, fires, earthquakes, volcanic eruptions, hurricanes, tornadoes, and other severe storms) may have positive or negative impacts on living things.	❍	❍	❍	★	✓		★	✓

★ **Standard Covered** ❍ **Standard to be covered** ✓ **Standard previously covered**

Correlation to the New York State Learning Standards and Major Understandings

This worktext is customized to the *New York Elementary Science Core Curriculum* and will help you prepare for the *Grade 4 Elementary-Level Science Test.*

After the lesson is completed, place a (✓) to indicate Mastery or a (**X**) to indicate Review Needed.

Chapter 6: Matter									
	Review Skill								
	Mastered Skill								
	Lessons	24	25	26	NA	NA	NA	Chap Review	End Rev
PS3.1	Observe and describe properties of materials, using appropriate tools.								
PS3.1a	Matter takes up space and has mass. Two objects cannot occupy the same place at the same time.	★	✓	✓				★	✓
PS3.1b	Matter has properties (color, hardness, odor, sound, taste, etc.) that can be observed through the senses.	★	✓	✓				★	✓
PS3.1c	Objects have properties that can be observed, described, and/or measured: length, width, volume, size, shape, mass or weight, temperature, texture, flexibility, reflectiveness of light.	★	✓	✓				★	✓
PS3.1d	Measurements can be made with standard metric units and nonstandard units.	★	✓	✓				★	✓
PS3.1e	The material(s) an object is made up of determine some specific properties of the object (sink/float, conductivity, magnetism). Properties can be observed or measured with tools such as hand lenses, metric rulers, thermometers, balances, magnets, circuit testers, and graduated cylinders.	★	★	✓				★	✓
PS3.1f	Objects and/or materials can be sorted or classified according to their properties.	❍	★	✓				★	✓
PS3.1g	Some properties of an object are dependent on the conditions of the present surroundings in which the object exists. For example: • temperature—hot or cold • lighting—shadows, color • moisture—wet or dry	❍	★	✓				★	✓
PS3.2	Describe chemical and physical changes, including changes in states of matter.								
PS3.2a	Matter exists in three states: solid, liquid, gas. • Solids have a definite shape and volume. • Liquids do not have a definite shape but have a definite volume. • Gases do not hold their shape or volume.	❍	❍	★				★	✓
PS3.2b	Temperature can affect the state of matter of a substance.	❍	❍	★				★	✓
PS3.2c	Changes in the properties or materials of objects can be observed and described.	❍	❍	★				★	✓
PS4.1	Describe a variety of forms of energy (e.g., heat, chemical, light) and the changes that occur in objects when they interact with those forms of energy.								
PS4.1f	Heat can be released in many ways, for example, by burning, rubbing (friction), or combining one substance with another.	❍	❍	★				★	✓

Chapter 7: Energy									
	Review Skill								
	Mastered Skill								
	Lessons	27	28	29	NA	NA	NA	Chap Review	End Rev
PS4.1	Describe a variety of forms of energy (e.g., heat, chemical, light) and the changes that occur in objects when they interact with those forms of energy.								
PS4.1a	Energy exists in various forms: heat, electric, sound, chemical, mechanical, light.	★	✓	✓				★	✓
PS4.1b	Energy can be transferred from one place to another.	❍	★	✓				★	✓
PS4.1c	Some materials transfer energy better than others (heat and electricity).	❍	★	★				★	✓

★ **Standard Covered** ❍ **Standard to be covered** ✓ **Standard previously covered**

Correlation to the New York State Learning Standards and Major Understandings

This worktext is customized to the *New York Elementary Science Core Curriculum* and will help you prepare for the *Grade 4 Elementary-Level Science Test.*

After the lesson is completed, place a (✓) to indicate Mastery or a (**X**) to indicate Review Needed.

Chapter 7: Energy (continued)								
Review Skill								
Mastered Skill								
Lessons	27	28	29	NA	NA	NA	Chap Review	End Rev
PS 4.1d Energy and matter interact: water is evaporated by the Sun's heat; a bulb is lighted by means of electrical current; a musical instrument is played to produce sound; dark colors may absorb light, light colors may reflect light.	❍	❍	★				★	✓
PS 4.1e Electricity travels in a closed circuit.	★	★	✓				★	✓
PS 4.1f Heat can be released in many ways; for example, by burning, rubbing (friction), or combining one substance with another.	★	✓	✓				★	✓
PS 4.1g Interactions with forms of energy can be either helpful or harmful.	★	✓	✓				★	✓
PS 4.2 Observe the way one form of energy is transferred into another form of energy present in common situations (e.g., mechanical to heat energy, mechanical to electrical energy, chemical to heat energy).								
PS 4.2a Everyday events involve one form of energy being changed to another. • Animals convert food to heat and motion. • The Sun's energy warms the air and water.	❍	★	✓				★	✓
PS 4.2b Humans utilize interactions between matter and energy. • chemical to electrical, light, and heat: battery and bulb • electrical to sound (e.g., doorbell buzzer) • mechanical to sound (e.g., musical instruments, clapping)o light to electrical (e.g., solar-powered calculator)	❍	❍	★				★	✓

Chapter 8: Force and Motion								
Review Skill								
Mastered Skill								
Lessons	30	31	32	NA	NA	NA	Chap Review	End Rev
PS5.1 Describe the effects of common forces (pushes and pulls) of objects, such as those caused by gravity, magnetism, and mechanical forces.								
PS5.1a The position of an object can be described by locating it relative to another object or the background (e.g., on top of, next to, over, under, etc.).	★	✓	✓				★	✓
PS5.1b The position or direction of motion of an object can be changed by pushing or pulling.	★	✓	✓				★	✓
PS5.1c The force of gravity pulls objects toward the center of Earth.	❍	❍	★				★	✓
PS5.1d The amount of change in the motion of an object is affected by friction.	★	✓	✓				★	✓
PS5.1e Magnetism is a force that may attract or repel certain materials.	❍	❍	★				★	✓
PS5.1f Mechanical energy may cause change in motion through the application of force and through the use of simple machines such as pulleys, levers, and inclined planes.	★	★	✓				★	✓
PS5.2 Describe how forces can operate across distances.								
PS5.2a The forces of gravity and magnetism can affect objects through gases, liquids, and solids.	❍	❍	★				★	✓
PS5.2b The force of magnetism on objects decreases as distance increases.	❍	❍	★				★	✓

★ **Standard Covered** ❍ **Standard to be covered** ✓ **Standard previously covered**

Correlation of the Performance Tasks to the New York State Learning Standards and Major Understandings

This worktext is customized to the *Elementary Science Core Curriculum* and will help you prepare for the *Grade 4 Elementary-Level Science Test.*

After the Performance Task is completed, place a (✓) to indicate Mastery or a (**X**) to indicate Review Needed.

Performance Task									
	Review Skill								
	Mastered Skill								
	Chapter	1	2	3	4	5	6	7	8
T1.1	**Engineering Design** Describe objects, imaginary or real, that can be modeled or made differently and suggest ways in which the objects can be changes, fixed or improved.								
T1.1a	Identify a simple/common object which might be improved and state the purpose of the improvement.	❍	❍	❍	❍	❍	❍	★	✓
T1.1b	Identify features of an object that help or hinder the performance of the object.	★	✓	✓	★	✓	✓	★	✓
T1.1c	Suggest ways the object can be made differently, fixed, or improved within given constraints.	❍	❍	❍	★	✓	✓	★	✓
T1.2	**Engineering Design** Investigate prior solutions and ideas from books, magazines, family, friends, neighbors, and community leaders.								
T1.2a	Identify appropriate questions to ask about the design of an object.	★	★	✓	✓	✓	✓	✓	✓
T1.2b	Identify the appropriate resources to use to find out about the design of an object.	❍	★	✓	✓	✓	✓	✓	✓
T1.2c	Describe prior designs of the object.	❍	❍	❍	★	✓	✓	✓	✓
T1.3	**Engineering Design** Generate ideas for possible solutions, individually and through group activity; apply age-appropriate mathematics and science skills; evaluate the ideas and determine the best solution; and explain reasons for the choices.								
T1.3a	List possible solutions, applying age-appropriate math and science skills.	❍	★	✓	✓	✓	✓	✓	✓
T1.3b	Develop and apply criteria to evaluate possible solutions.	❍	★	✓	✓	✓	★	✓	✓
T1.3c	Select a solution consistent with given constraints and explain why it was chosen.	❍	★	✓	✓	✓	★	★	✓
T1.4	**Engineering Design** Plan and build, under supervision, a model of a solution, using familiar materials, processes, and hand tools.								
T1.4a	Create a grade-appropriate graphic or plan listing all materials needed, showing sizes of parts, indicating how things will fit together, and detailing steps for assembly.	❍	❍	❍	❍	❍	❍	❍	★
T1.4b	Build a model of the object, modifying the plan as necessary.	❍	❍	❍	❍	❍	❍	❍	★
T1.5	**Engineering Design** Discuss how best to test the solution; perform the test under teacher supervision; record and portray results through numerical and graphic means; discuss orally why things worked or didn't work; and summarize results in writing, suggesting ways to make a solution better.								
T1.5a	Determine a way to test the finished solution or model.	❍	❍	★	✓	✓	✓	✓	★
T1.5b	Perform the test and record the results, numerically and /or graphically	❍	❍	★	✓	✓	✓	★	★
T1.5c	Analyze results and suggest how to improve the solution or model, using oral, graphic, or written formats.	❍	❍	★	✓	✓	✓	★	★
6.1	**Interconnectedness: Common Themes** Through systems thinking, people can recognize the commonalities that exist among all systems and how parts of a system interrelate and combine to perform specific functions. • Observe and describe interactions among components of simple systems. • Identify common things that can be considered to be systems (e.g., a plant, a transportation system, human beings).	❍	❍	❍	★	✓	✓	✓	★

★ **Standard Covered** ❍ **Standard to be covered** ✓ **Standard previously covered**

Correlation of the Performance Tasks to the New York State Learning Standards and Major Understandings

This worktext is customized to the *Elementary Science Core Curriculum* and will help you prepare for the *Grade 4 Elementary-Level Science Test.*

After the Performance Task is completed, place a (✓) to indicate Mastery or a (**X**) to indicate Review Needed.

Performance Task (Continued)								
Review Skill								
Mastered Skill								
Chapter	1	2	3	4	5	6	7	8
6.2 Interconnectedness: Common Themes Models are simplified representations of objects, structures, or systems, used in analysis, explanation, or design. • Analyze, construct, and operate models in order to discover attributes of the real thing. • Discover that a model of something is different from the real thing but can be used to study the real thing. • Use different types of models, such as graphs, sketches, diagrams, and maps, to represent various aspects of the real world.	❍	❍	❍	★	✓	✓	✓	✓
6.3 Interconnectedness: Common Themes The grouping of magnitudes of size, time, frequency, and pressures or other units of measurement into a series of relative order provided a useful way to deal with the immense range and the changes in scale that affect behavior and design or systems • Observe that things in nature and things that people make have very different sizes, weights, and ages. • Recognize that almost anything has limits on how big or small it can be.	❍	❍	❍	❍	★	✓	✓	✓
6.4 Interconnectedness: Common Themes Equilibrium is a state of stability due either to a lack of changes (static equilibrium) or a balance between opposing forces (dynamic equilibrium). • Observe that things change in some ways and stay the same in some ways. • Recognize that things can change in different ways such as size, weight, color, and movement. Some changes can be detected by taking measurements.	❍	❍	❍	★	✓	✓	✓	✓
6.5 Interconnectedness: Common Themes Identifying patterns of change is necessary for making predictions about future behavior and conditions. • Use simple instruments to measure such quantities as distance, size, weight and look for patterns in the data. • Analyze data by making tables and graphs and looking for patterns of change.	❍	❍	★	✓	✓	✓	✓	✓
6.6 Interconnectedness: Common Themes In order to arrive at the best solution that meets criteria within given constraints, it is often necessary to make trade-offs. • Choose the best alternative of a set of solution under given constraints. • Explain the criteria used in selecting a solution orally and in writing.	❍	★	✓	✓	✓	✓	✓	✓
7.1 Interdisciplinary Problem Solving The knowledge and skills of mathematics, science, and technology are used together to make informed decisions and solve problems, especially those relating to issues of science/technology/society, consumer decision making, design, and inquiry into phenomena. • Analyze science/technology/society problems and issues that affect their home, school, or community, and carry out a redial course of action. • Make informed consumer decisions by applying knowledge about the attributes of particular products and making cost/benefit trade-offs to arrive at an optimal choice. • Design solutions to problems involving a familiar and real context, investigate related science concepts to determine the solution, and use mathematics to model, quantify, measure, and compute . • Observe phenomena and evaluate them scientifically and mathematically by conducting a fair test of the effect of variables and using mathematical knowledge and technological tools to collect, analyze, and present data and conclusions.	❍	❍	★	✓	✓	✓	✓	★
7.2 Interdisciplinary Problem Solving Solving interdisciplinary problems involves a variety of skills and strategies, including effective work habits; gathering and processing information; generating and analyzing ideas; realizing ideas; making connections among common themes of mathematics, science, and technology; and presenting results. • work effectively • gather and process information • generate and analyze ideas • observe common themes • realize ideas • present results	★	★	★	★	★	★	★	★

★ **Standard Covered** ❍ **Standard to be covered** ✓ **Standard previously covered**

To the Student,

How do you get better at anything you do? You practice! Just like with sports or other activities, the key to success in school is practice, practice, practice.

This book will help you review and practice science strategies and skills. These are the strategies and skills you need to know to measure up to the New York State Learning Standards for Science for your grade. Practicing these skills and strategies now will help you do better in your work all year.

This Measuring Up® book has 8 chapters. Chapter one provides review and practice with scientific processes. Chapters two through four provides review and practice with topics about the living environment. Chapters five through eight provides practice with topics about the physical setting.

Each lesson consists of three main sections:

- **Focus on New York State Learning Standards** introduces the skills covered in the lesson.
- **Guided Instruction** guides you through a review of science concepts and skills you will need for successful learning.
- **Apply the New York State Learning Standards to the State Test** gives you practice in answering the types of multiple-choice and open-ended questions you will see on the *Grade 4 Elementary-Level Science Test.*

In addition to the lessons, other sections in the book are called **Building Stamina®**. These sections contain multiple-choice and open–ended questions that help build your intellectual brain power. Many of these questions are more difficult and will help you prepare for taking tests.

There is also a section in each chapter called **Higher-Order Performance Task**. In this section, you will be performing short experiments, based on skills you have learned throughout that particular chapter. **Higher-Order Performance Tasks** are designed to tap your higher-order thinking skills.

This year, you will take the *Grade 4 Elementary-Level Science Test*. It will be an important step forward. This test will show how well you measure up to the New York State Learning Standards. It is just one of the many important tests you will take. Success on this test will prepare you for the next level of science challenges. Have a great and successful year!

To Parents and Families,

All students need science skills to succeed. New York educators have created the New York State Learning Standards for Science. The standards describe what all New York students should know at each grade level. Students need to meet these standards to graduate, as measured by the *Grade 4 Elementary-Level Science Test* given later this year.

The test is directly related to the New York State Learning Standards. The test emphasizes higher-order thinking skills. Students must learn to consider, analyze, interpret, and evaluate instead of just recalling simple facts.

Measuring Up® will help your child to review the learning standards and prepare for all science exams. It contains:

- **Lessons** that focus on practicing the New York State Learning Standards for Science.
- **Guided Instruction** in which students are shown the steps and skills necessary to solve a variety of science problems.
- **Apply the New York State Learning Standards to the State Test**, which shows how individual standards can be understood through multiple-choice and open-ended questions, similar to those on the grade 4 science test.
- **Building Stamina®**, which gives practice with more difficult multiple-choice and open–ended questions that require higher-level thinking.
- **Higher-Order Performance Tasks,** which give practice in performing classroom experiments using short-response questions and data tables, and creating drawings.

For success in school and the real world, your child needs to be successful in science. Get involved! Your involvement is crucial to your children's success. Here are some suggestions:

- Make sure your home shows that science is important. Involve your child in activities that require science concepts and skills, such as mixing recipes, exploring the ecology of your neighborhood and community, observing and studying the night sky, and recycling.
- Help to find appropriate Internet sites for science.
- Note how science is used when you are out with your family. Discuss how science is used in preparing meals, in careers such as medicine and architecture, in space exploration, and in other real-life applications.
- Encourage your child to talk about what he or she has learned in science class.
- Encourage your child to take time to review and check his or her homework.

Work this year to ensure your child's success. Science skills are essential skills for success and pleasure throughout your child's life.

New York State Learning Standards
and Success Strategies for the State Test

SAFETY FIRST

This book contains various investigations and activities that demonstrate the concepts in Measuring Up® *to the New York State Learning Standards and Success Strategies for the State Test*. Following standard safety practices is an important laboratory procedure when completing any science activity.

Before You Experiment

1. Read the instructions for each science activity before you begin. Review the safety symbols and instructions so that special safety equipment is ready before you start.
2. Wash all the tools you will use.
3. Make sure your teacher or another adult is present to supervise your work.
4. Wear the safety equipment that your teacher tells you to. If your hair is long, be sure to tie your hair back.

During an Experiment

5. Follow the instructions step-by-step in the order that they are presented.
6. Never run in a lab or play games during an experiment.
7. Do not bring food or drink into the lab or classroom.
8. Substances used in experiments can be dangerous. Only taste them or smell them if your teacher tells you to.
9. Mix ingredients only as your activity instructs. Playing with these ingredients may create dangerous or explosive substances.
10. Remember, knives and scissors are sharp. Move the knife or scissors away from your body when you are cutting.
11. Accidents do occur. Someone may be hurt or something may be broken. Immediately tell your teacher or the adult supervising your work.
12. Check to see that all containers are labeled so you know what substances they hold.

After the Experiment Is Done

13. Ask your teacher what to do with unused ingredients and containers.
14. Follow your teacher's instructions to clean up your work area.
15. Make sure you turn off all lights, switches, burners, and faucets.

COMMON SAFETY EQUIPMENT SYMBOLS

Safety goggles must be worn	Use electrical safety guidelines	Animal safety
Gloves must be worn	Do not use an open flame	Sharp object
Protective clothing must be worn	Poisonous chemicals	Plant safety

What's Ahead in Measuring Up®

What's Ahead in Measuring Up®

This book was created for New York students like you. Each lesson and question and every investigation is aimed at helping you master the New York State Learning Standards for Science and do well on the *Grade 4 Elementary-Level Science Test*. It will also help you do well on other science exams you take during the school year.

About the Test

New York educators have set up standards for science. They are called the New York State Learning Standards for Science. They spell out what all students at each grade level should know. New York educators have also created a state-wide test for science. It is called the *Grade 4 Elementary-Level Science Test*. It shows how well students have mastered the learning standards. Test questions go along with and meet the following learning standards:

Science Standards

STANDARD 1

Analysis, Inquiry, and Design Students will use mathematical analysis, scientific inquiry, and engineering design, as appropriate, to pose questions, seek answers, and develop solutions.

STANDARD 2

Information Systems Students will access, generate, process, and transfer information using appropriate technologies.

STANDARD 4

The Physical Setting and the Living Environment Students will understand and apply scientific concepts, principles, and theories pertaining to the physical setting and living environment and recognize the historical development of ideas in science.

STANDARD 6

Interconnectedness: Common Themes Students will understand the relationships and common themes that connect mathematics, science, and technology and apply the themes to these and other areas of learning.

STANDARD 7

Interdisciplinary Problem Solving Students will understand the relationships and common themes that connect mathematics, science, and technology to address real-life problems and make informed decisions.

Format of the Test

To help you prepare for the *Grade 4 Elementary-Level Science Test*, Measuring Up® includes:

- Multiple-choice questions
- Constructed response questions
- Performance Tasks

Many questions include a picture, diagram, chart, or other type of graphic, which is used to answer the question. Measuring Up® gives you practice in reading and using these types of graphics.

What's Ahead in Measuring Up®

Measuring Up® on Multiple-Choice Questions

You are probably familiar with the multiple-choice type of question. It has a question, or stem, followed by four answer choices. Your job is to select the one correct choice. On the New York science test, you will answer many multiple-choice questions. Here are some tips:

- Always try to determine the answer without looking at the choices. Once you arrive at an answer, compare your response with the choices.
- If your answer is not one of the choices, check your work. Rethink your ideas carefully because the answer may be one of the choices and an error may have led you astray.
- Some multiple-choice questions refer to a graphic such as an illustration, a graph, a table, or a picture. You will be asked to read or interpret the graphic. Read the question carefully and use the graphic to answer the specific question.
- Many questions test higher-order thinking skills. You must connect the ideas and information to come up with the right answers.
- Even if you don't know the answer to a multiple-choice question, you can make a good guess based on what you know and get the question right.
- Check and double-check your answers before you turn in the test. Be sure of your answers and be sure you haven't marked a wrong answer choice by mistake.

Measuring Up® on Open-Ended Questions

Open-ended items usually require you to study a graphic and sometimes read a short paragraph, too. Then you will write short responses to questions related to the graphic or paragraph. Here are some tips:

- Carefully study the graphic or the paragraphs. Because you do not have answer choices as a way to check yourself, it is important to take your time and follow all the steps carefully.
- Once you have an answer, carefully write it into the space provided.
- After completing the test, look back at your response to make sure you have responded directly to the questions.

Measuring Up® on Higher-Order Performance Tasks

At the end of each chapter is a Higher-Order Performance Task, which will require you to use science skills. You will enjoy performing these tasks because you will be able to use science equipment to help you answer the questions. These Performance Tasks are new and are designed to assess higher-order thinking skills. Here are some tips:

- Follow your teacher's directions.
- Read the Performance Task carefully.
- Gather the materials needed.
- Follow all safety guidelines.
- Follow each step in order.
- Answer all the questions.
- Return all the materials the way you found them.
- Check your answers.

Measuring Up® with **Building Stamina®**

A unique feature of Measuring Up® is **Building Stamina®**, designed to give you practice and build your confidence and endurance for completing higher level thinking activities. These activities include answering questions that cover multiple learning standards. Each chapter ends with a **Building Stamina®** section. At the end of the book is a longer, comprehensive **Building Stamina®**, which is a complete review of all the standards covered in the lessons.

Higher-Order Thinking Skills

The *Grade 4 Elementary-Level Science Test* is designed to tap your higher-order thinking skills. When you use higher-order thinking skills, you do more than just recall information. For instance, instead of being asked to name the parts of the circulatory system, you might need to describe how this system functions and affects the body of a living organism. Instead of being asked to identify the components of an ecosystem, you might have to predict how certain changes might affect the ecosystem.

NYS Test Tips

Throughout this book, you will find little boxes that look like this:

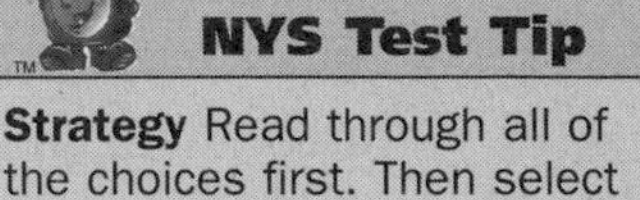

Strategy Read through all of the choices first. Then select the best answer.

These NYS Test Tips help you think about the test and focus on certain lesson skills. Here are some other things you should keep in mind, both on and before test day:

- Start preparing now. Pace yourself. Spend a few minutes a day practicing answering test questions. Right now, the test may seem far in the future, but you will be sitting with a test booklet in front of you before you know it.
- Get a good night's sleep the night before the test. Do not expect to cram everything into your head the night before. You can't remember much that way, and you will be too tired to do well.
- Eat a good breakfast. If you are hungry during the test, you will be distracted and unable to think clearly.
- Think positively. Do not focus on the things you do not know. Nobody is perfect. If you are unsure of an answer, mark that question and move on to the next. After you have worked through the test, return to those questions you have marked.

You will learn a lot in Measuring Up®. You will review and practice the New York State Learning Standards for Science. You will practice for the *Grade 4 Elementary-Level Science Test*. Finally, you will build your stamina to answer tough questions. You will more than measure up. You'll be a smashing success!

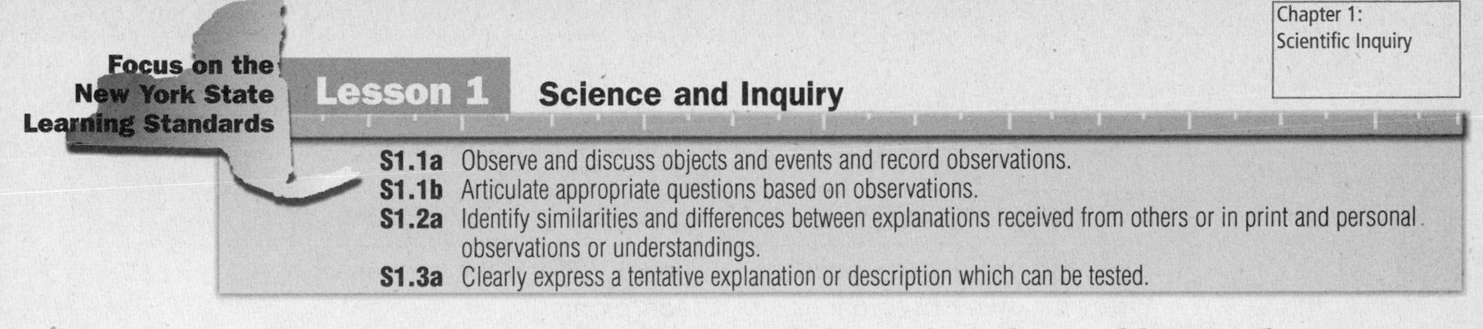

Lesson 1 Science and Inquiry

S1.1a Observe and discuss objects and events and record observations.
S1.1b Articulate appropriate questions based on observations.
S1.2a Identify similarities and differences between explanations received from others or in print and personal observations or understandings.
S1.3a Clearly express a tentative explanation or description which can be tested.

You can use scientific inquiry to help explain the world around you.

A **scientist** is a person who has knowledge of science that is based on observed facts and tested truths.

Scientific inquiry is the way scientists find facts and solve problems.

Guided Instruction

Directions Read the following information.

People have always wondered about the world and universe. **Scientists** search for facts and explanations about our world and universe. They ask questions of why and how in order to have a greater understanding about objects and events. To answer questions, they use the methods of **scientific inquiry**.

Scientists make observations using their senses of sight, hearing, touch, smell, and/or taste. They look to the work of others to find explanations that are the same or different from their own possible explanations. They set up experiments to test their own explanations. Everyone can be a scientist when they make observations. How can you use your senses to study a rabbit?

Guided Questions

Can anyone be a **scientist?**

What is **scientific inquiry?**

What senses help you to make observations?

What question can you ask about a rabbit based on your observations?

Look at the chart below to see how scientific inquiry can be used to explain the relationship of sunlight and plant growth.

STEPS IN SCIENTIFIC INQUIRY	EXAMPLE
1. State the problem or question. Use your own observations, using your senses.	Sunlight affects the growth of bean plants.
2. Research information. What have others discovered about the problem?	You have seen that bean plants grow from seeds. You learn that plants need the right amount of light, water, air, and temperature to grow.
3. Form an explanation, or a best guess, that can be tested by doing an experiment.	Too much light affects the growth of bean plants. Too little light affects the growth of bean plants.

Directions For each question, write your answer in the space provided.

1. How do scientists find facts and solve problems?

____

2. How many of your senses can help you make scientific observations?

____

3. What is the first step in scientific inquiry?

____

____

4. What is the next step after an explanation, or best guess, has been formed?

____

5. Does the experiment always prove a scientist's explanation to be true? Why or why not?

____

____

6. A scientist placed two bean plants in bright sunlight, two bean plants in shade, and two bean plants in a dark closet. Write a question that she is most likely trying to answer.

____

____

Apply the New York State Learning Standards to the State Test

Directions: Read the paragraph and look at the chart below. Use this information to fill in the second column of the scientific inquiry chart, 7 through 9.

You wonder how air temperature will affect the air inside a balloon. You know that when cold air is warmed, it expands. You have seen hot air balloons in which the air inside the balloon is heated to make the balloon, basket, and pilot rise into the air. In science class, you learned that cold air is denser than hot air.

You blow up a balloon and tie the end. First you place the balloon in a refrigerator for a while. Then you place the balloon in direct sunlight for a while.

STEPS IN SCIENTIFIC INQUIRY	YOUR SCIENTIFIC INQUIRY
State the problem or question. Use your own observations, using your senses.	7.
Research information. What have others discovered about the problem?	8.
Form an explanation, or a best guess, that can be tested by doing an experiment.	9.

Directions (10–15): Each question is followed by four choices. Decide which choice is the best answer. Circle the letter of the answer you have chosen.

10 What is a person who has knowledge based on observed facts and tested truths called?

A a professor
B balloonist
C a scientist
D an experimenter

11 Which is the first step in scientific inquiry?

A Fill out a scientific inquiry chart.
B State a problem or question.
C Research information about your question.
D Look at the results of your experiment.

12 Which is the second step in scientific inquiry?

A Fill out a scientific inquiry chart.
B Ask questions.
C Research information about your question.
D Form an explanation that can be tested.

13 Which is the third step in scientific inquiry?

A Fill out a scientific inquiry chart.
B Form an explanation that can be tested.
C Research information about your question.
D Ask questions.

14 Juan does an experiment. He drops a Ping-Pong ball, a golf ball, and a tennis ball from the same height. Which of the following explanations is he most likely testing?

A The golf ball is hardest of the three balls.
B The Ping-Pong ball is lightest of the three balls.
C The golf ball bounces highest of the three balls.
D The tennis ball is largest of the three balls.

15 A class is observing how pendulums move. The students built the three pendulums shown below. Which question are they asking by using these pendulums?

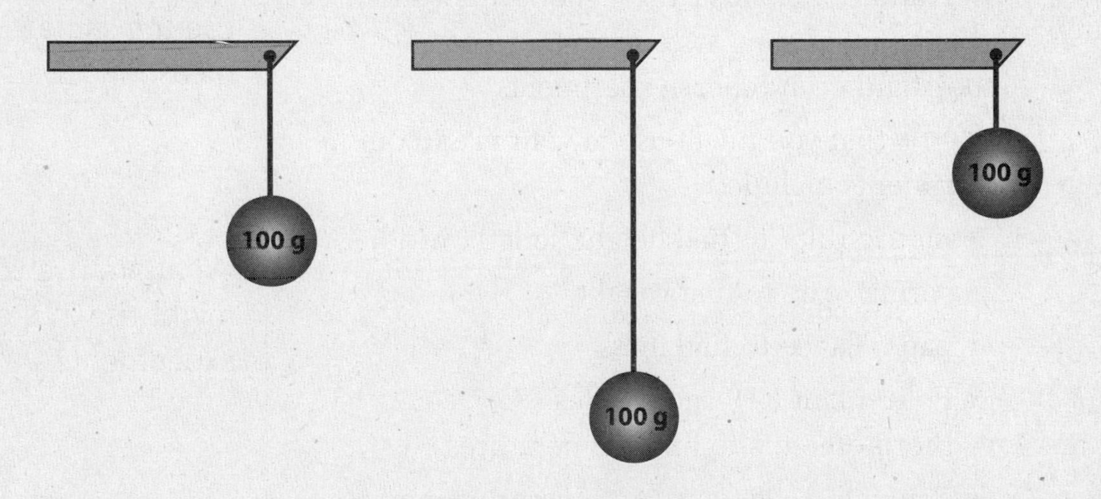

A How does the length of the string affect the motion?
B How does the color of the string affect the motion?
C How does the shape of the weight affect the motion?
D How does the thickness of the string affect the motion?

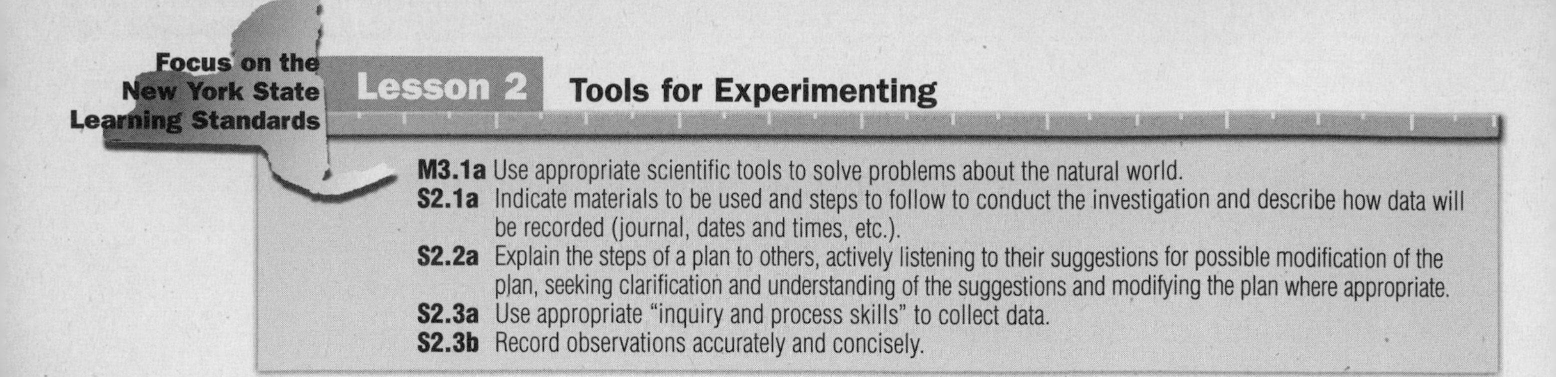

Lesson 2 Tools for Experimenting

M3.1a Use appropriate scientific tools to solve problems about the natural world.
S2.1a Indicate materials to be used and steps to follow to conduct the investigation and describe how data will be recorded (journal, dates and times, etc.).
S2.2a Explain the steps of a plan to others, actively listening to their suggestions for possible modification of the plan, seeking clarification and understanding of the suggestions and modifying the plan where appropriate.
S2.3a Use appropriate "inquiry and process skills" to collect data.
S2.3b Record observations accurately and concisely.

You can use the correct tools to make scientific observations.

Tools are the instruments that help you conduct an experiment.
Recording observations means to write down the information gathered during an experiment.

Directions Read the following information.

Scientists use instruments to help them make observations. Scientists throughout the world use the same measuring systems so they can understand each other's experiments.

Tools that you might use in your classroom experiments include:

- metric ruler to find height, length, width
- spring scale to find weight
- pan balance to find mass
- Fahrenheit (°F) and Celsius (°C) thermometers to find temperatures
- graduated cylinders and measuring cups to find volume

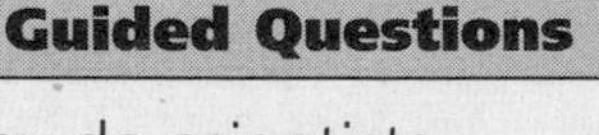

Why do scientists around the world use the same measuring systems?

What are five **tools** you might use in a classroom experiment?

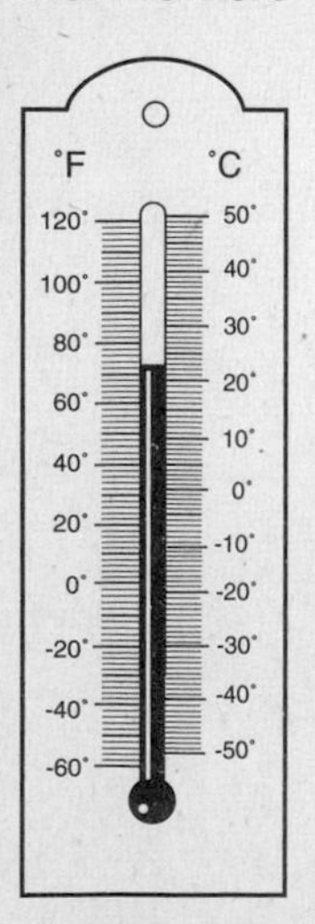

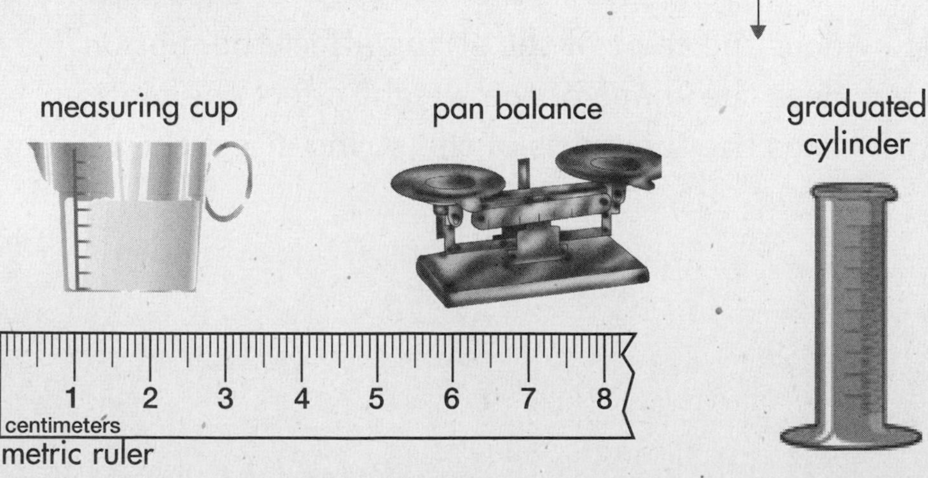

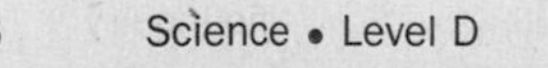

Scientists also plot their data, that is, the information gathered from their observations, into graphs. Putting data into graphs helps to show patterns. Finding patterns helps scientists interpret, or understand, their data. A scientist who observed the daily temperature of an object recorded his data on a graph like the one below. Notice that on the left side of the graph are the possible temperatures in degrees Celsius (°C). Along the bottom of the graph are the days of the week the temperatures were observed. A dot is placed on the day line opposite of the temperature observed on that day. Then the dots are connected to show a pattern.

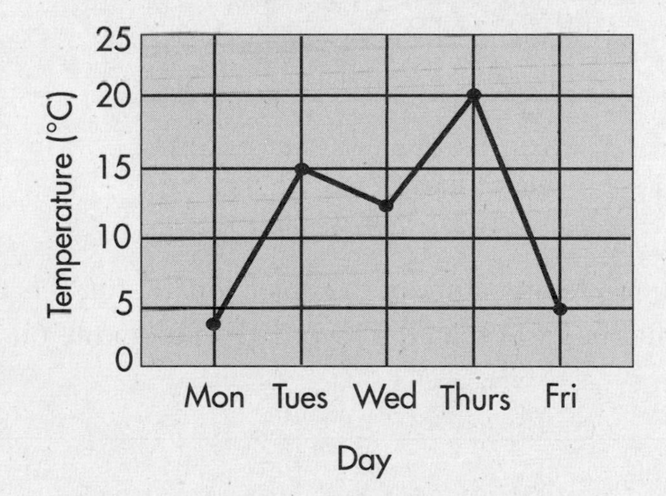

Scientists also keep journals to **record observations**, the materials they used in the experiment, and the steps they followed. This helps the scientist explain all of the details and procedures of the experiment to other scientists. The recorded information helps the other scientists repeat the experiment, so they can check the results themselves.

Guided Questions

What do graphs help to show about data?

Name two ways to **record observations.**

Directions For each question, write your answer in the space provided.

1. Which tools will you use to measure the height, weight, mass, and volume of a cube of metal?

height ______________________________

weight ______________________________

mass ______________________________

volume ______________________________

Base your answers to questions 2 and 3 on the graph below.

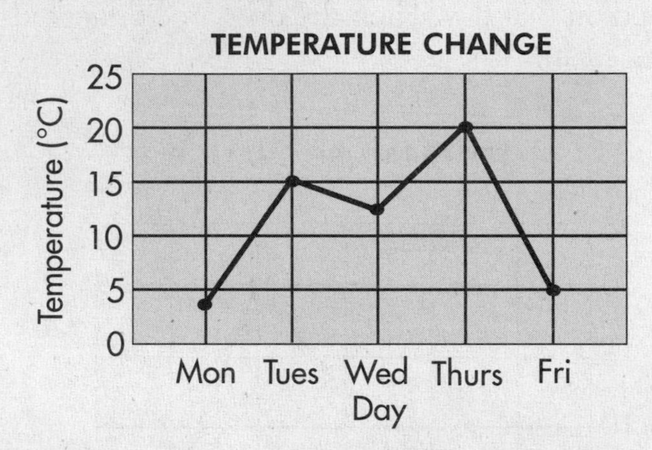

2. What is the recorded temperature of the object on Tuesday? Is this temperature higher or lower than the recorded temperature on Monday?

____

____

3. On what day was the temperature recorded at exactly 5°C?

____

4. Why is it important for scientists to keep accurate records?

____

____

Apply the New York State Learning Standards to the State Test

Directions: For each question, write your answer in the space provided. Base your answers to questions 5 through 8 on the paragraph, chart, and graph below.

You are wondering if the outdoor temperature at noon is the same or different each day. You decide to make a few observations. Using a Celsius thermometer, you take the following measurements for three days.

DAY	NOON TEMPERATURE
Monday	20°C
Tuesday	24°C
Wednesday	26°C

5 What tool did you use to make your observations?

____

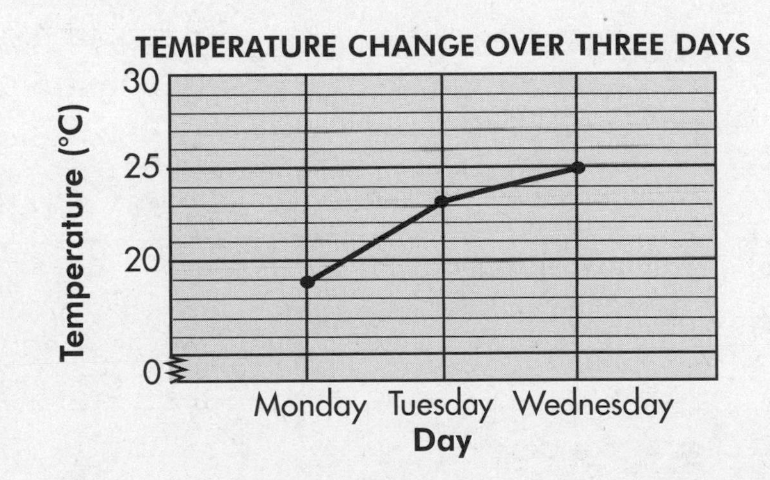

6 Use the data in the chart above to plot the data on the graph. Connect the dots you plot to show a pattern.

7 Is the temperature the same or different each day of observation?

____

8 What pattern do you see in your data?

____

Directions (9–14): Each question is followed by four choices. Decide which choice is the best answer. Circle the letter of the answer you have chosen.

Base your answers to questions 9 and 10 on the drawing below.

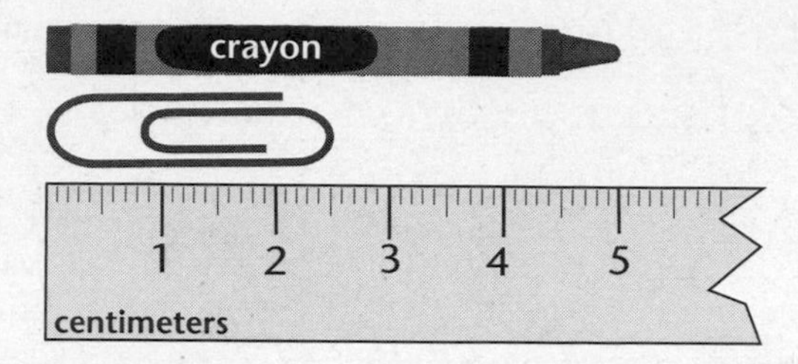

9 What is the science tool shown above measuring?

A mass
B width
C length
D temperature

10 What is the length of the crayon to the nearest centimeter?

A about 3 cm
B about 4 cm
C about 5 cm
D about 6 cm

11 What is best measured with the tool shown below?

A color of apple skin
B length of an apple stem
C height of an apple tree
D volume of apple juice

12 Why is a journal useful to a scientist?

A to write letters home
B to write short stories
C to list the names of everyone who helped in an experiment
D to record observations made in an experiment

13 Which three tools would help you find the height, weight, and temperature of a horse?

A metric ruler, large spring scale, thermometer
B pan balance, large spring scale, thermometer
C metric ruler, thermometer, graduated cylinder
D metric ruler, pan balance, graduated cylinder

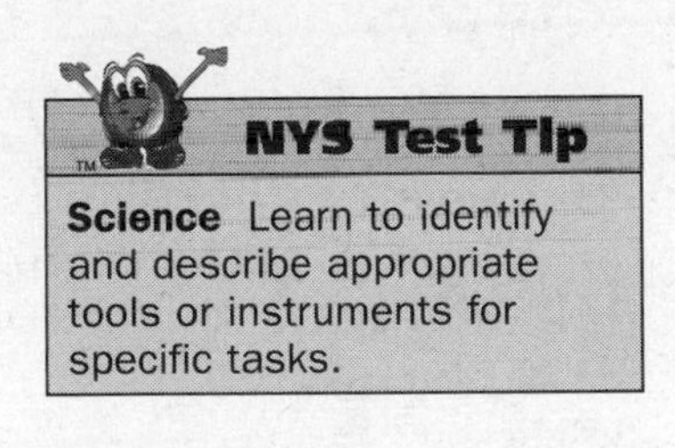

14 Look at the graph. What temperature is recorded for Thursday?

TEMPERATURE CHANGE

Temperature (°C)
25
20
15
10
5
0
Mon Tues Wed Thurs Fri
Day

A 5°C
B 10°C
C 13°F
D 15°C

Focus on the New York State Learning Standards

Lesson 3 Lab Safety

S2.1a Indicate materials to be used and steps to follow to conduct the investigation and describe how data will be recorded (journal, dates and times, etc.).
S2.3a Use appropriate "inquiry and process skills" to collect data.

During a science investigation, it is important to use safety procedures and to know the proper way to dispose of materials.

Safety guidelines are rules to follow when doing a science investigation. A **precaution** is something done to prevent an accident.

Guided Instruction

Directions Read the following information.

Science investigations can involve special tools and materials. To protect yourself and others, you must follow **safety guidelines**.

The most important safety guideline is to tell your teacher about accidents that happen. If a glass breaks, or something spills—even water—you need to tell your teacher about it. Your teacher is concerned about your safety and wants to know what happens in the classroom.

Before beginning a science investigation, read the instructions and ask questions about anything you do not understand.

Keep your work area clean and neat. A clean workspace can help to prevent spills and accidents. Then, take **precautions** to avoid mishaps. Safety goggles, gloves, and aprons can protect you from spills. Tie back long hair and remove loose clothing or jewelry that could touch flames or chemicals.

Guided Questions

What is the most important **safety guideline?**

What **precautions** has the student in the picture taken?

Once you begin an experiment, follow the steps. Do not change the steps without asking your teacher. If you aren't sure what to do or how to follow the steps, ask your teacher. If there are any accidents, report them to your teacher immediately. While you are working, do not put your hands or anything else into your mouth.

After you finish an experiment, you can still take safety precautions. Clean up your workspace. Clean your tools and put them away. If there are materials left over, ask your teacher what to do with them. Then, before you leave the room, wash your hands.

Guided Questions

What precautions can you take after you finish an experiment?

SAFETY GUIDELINES
1. Tell your teacher immediately about accidents that happen.
2. Before beginning a science investigation, read the instructions carefully.
3. Keep your work area clean and neat.
4. Wear safety goggles, gloves, and aprons.
5. Tie back long hair and remove loose clothing or jewelry.
6. Do not put your hands or anything else into your mouth.
7. Clean up your workspace. Clean your tools and put them away.
8. Before you leave the room, wash your hands.

Directions For each question, write your answer in the space provided.

1. Why should you follow safety guidelines?

____

2. What is the first thing that you should do before you start a science investigation?

____

3. What should you do if you break a glass bottle during an investigation?

____

4. What can you wear as a precaution against splashes or spills?

____

5. Which do you think is the most important safety guideline? Explain.

____

____

6. Why is it important to wash your hands after an experiment?

____

Apply the New York State Learning Standards to the State Test

Directions: For each question, write your answer in the space provided. Base your answers to questions 7 through 11 on the drawing and paragraph below.

Your class is getting ready for a science experiment. The class has been given instructions to read before the activity begins. Before you start the experiment, the class is given time to ask the teacher questions about the activity. Several classmates ask questions. During the experiment, you carry bottles of pond water. You also use rulers for measuring and tools for picking objects out of the water. Everyone wears safety goggles, gloves, and lab aprons. When your friend Eric breaks a bottle, he picks up the pieces of the bottle without telling anyone.

7 What are some dangers in this investigation?

____

____

8 What mistake does Eric make?

____

____

9 What are some precautions that are taken during this investigation?

____

____

____

10 After the investigation is over, students clean their tools and throw away leftover materials. What is left to do?

____

____

11 Why is it important to read the instructions before beginning an experiment?

____

____

Directions (12–17): Each question is followed by four choices. Decide which choice is the best answer. Circle the letter of the answer you have chosen.

12 What is the first thing you should do if something is spilled during a lab investigation?

A Wipe it up with a paper towel.
B Tell your teacher.
C Pour water on it.
D Ignore the spill.

13 Which of the following is a precaution you should take after a lab investigation is complete?

A Read the activity and ask questions.
B Wash your hands.
C Put on an apron and goggles.
D Take good notes.

14 Which of the following is the first thing you should do before an investigation begins?

A Make a data table.
B Clean up the equipment from the day before.
C Read over the investigation.
D Get out the protective equipment.

15 Which of the following is not worn as a safety precaution?

A 

B 

C 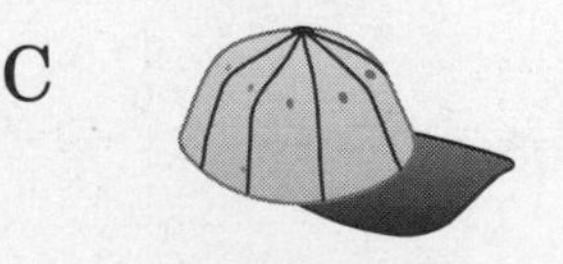

D 

16 When should you remove your safety goggles?

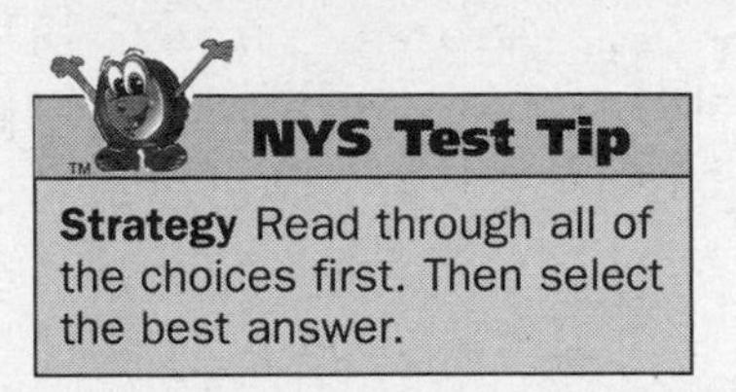

A after tools and materials are put away
B whenever you leave your seat
C while you talk to the teacher
D whenever you want to

17 What is the first thing you should do if something gets in your eye during an investigation?

A Rub your eye.
B Tell your teacher.
C Ask your partner to take it out.
D Pour water on it.

Focus on the New York State Learning Standards

Lesson 4 Analyze and Display Data

S3.1a Accurately transfer data from a science journal or notes to appropriate graphic organizer.
S3.2a State, orally and in writing, any inferences or generalizations indicated by the data collected.
S3.3a Explain their findings to others, and actively listen to suggestions for possible interpretations and ideas.
S3.4a State, orally and in writing, any inferences or generalizations indicated by the data, with appropriate modifications of their original prediction/explanation.
S3.4b State, orally and in writing, any new questions that arise from their investigation.

You can make inferences or generalizations by analyzing and displaying scientific data.

Data is information gathered in an experiment.

A **graphic organizer** is a drawing, chart, or graph, in which you can accurately display data.

Guided Instruction

Directions Read the following information.

When you do an experiment to try to answer a scientific question, you make observations and collect information. The information you gather from the experiment is your **data**. But what will you do with the data you have collected?

Scientists transfer, or move, their data from their science journal or notes into a **graphic organizer**. Here are three ways you can organize your data.

You might make a drawing of something you observed. For example, if you were observing the birds in your neighborhood, you might draw the birds you see.

A second way you might organize your observations is in a chart. For example, if you were observing the growth of two plants, you might put your data into a chart like this:

	PLANT 1	PLANT 2
Height at 1 week	1 cm	1 cm
Height at 2 weeks	5 cm	3 cm

A third way to organize data is by using a graph. A graph is useful for showing changes that occur during an experiment. A graph has lines that

Guided Questions

What is **data?**

What is one way to organize data?

cross, or intersect. The data is plotted onto the graph. For example, changes in daily temperatures can be shown on a graph.

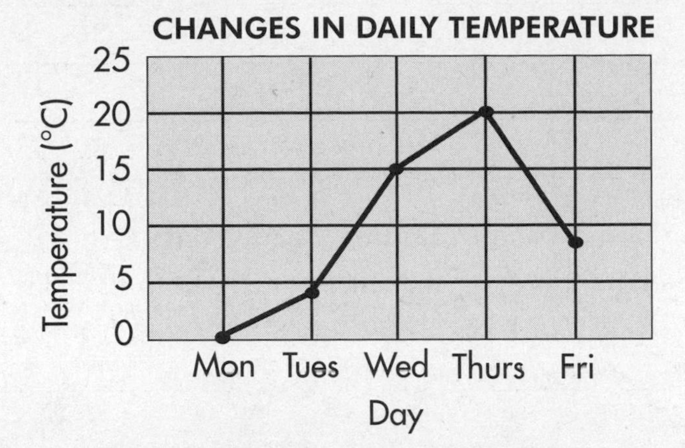

Organizing and displaying your observations in drawings, charts, and graphs makes studying the data easier. It helps you see your information or see patterns in your information. It helps you draw conclusions and write a report about the results of your experiment. Once you have drawn conclusions, your drawings, charts, or graphs will help you tell others about your experiment. Your data might lead you to new questions. For example, you might wonder if the pattern you see will continue. Or you might wonder if the data will change if you add or take away something in your experiment.

Guided Questions

What are three ways to record data?

How does a display of **data** help you draw conclusions?

Directions For each question, write your answer in the space provided.

1. If you are studying the shape of leaves on different trees, how would you organize your data? Why?

____

____

2. If you are studying the changes in water levels in a lake, would you put your data in a drawing or a graph? Why?

____

SUNSET TIMES

DAY	TIME
Monday	5:00 P.M.
Tuesday	5:02 P.M.
Wednesday	5:04 P.M.
Thursday	5:06 P.M.

3. The chart above shows sunset times for four days.
Based on the chart, predict what time sunset be on Friday.

____

4. What conclusion can you draw from the information in the sunset chart above?

____

Apply the New York State Learning Standards to the State Test

Directions: For each question, write your answer in the space provided. Base your answers to questions 5 through 8 on the information and data below.

The air temperatures below were taken at 3 P.M. on Monday, Tuesday, Wednesday, Thursday, and Friday.

Monday 8°C   Tuesday 13°C   Wednesday 18°C
Thursday 15°C   Friday 20°C

5 Complete the following chart with the data from above.

AIR TEMPERATURES AT 3 P.M.

DAY OF THE WEEK	TEMPERATURE

6 Which graph correctly displays the data from the chart in exercise 5 on page 20?

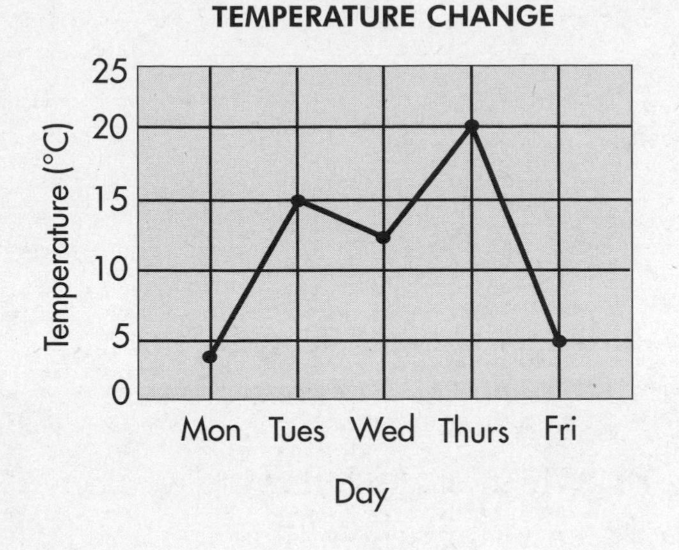

A

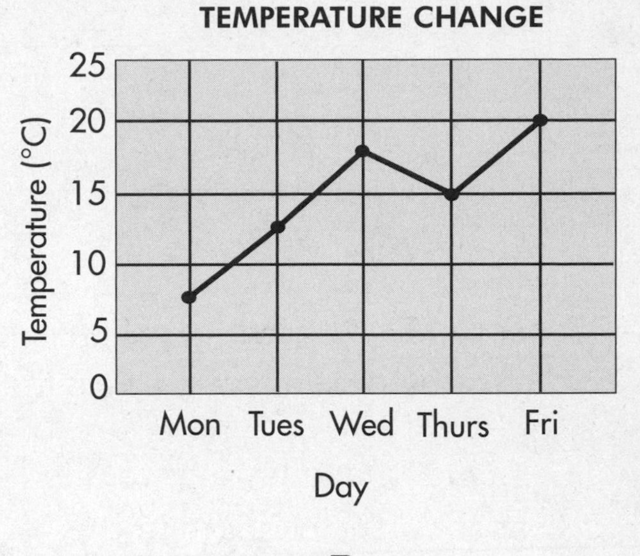

B

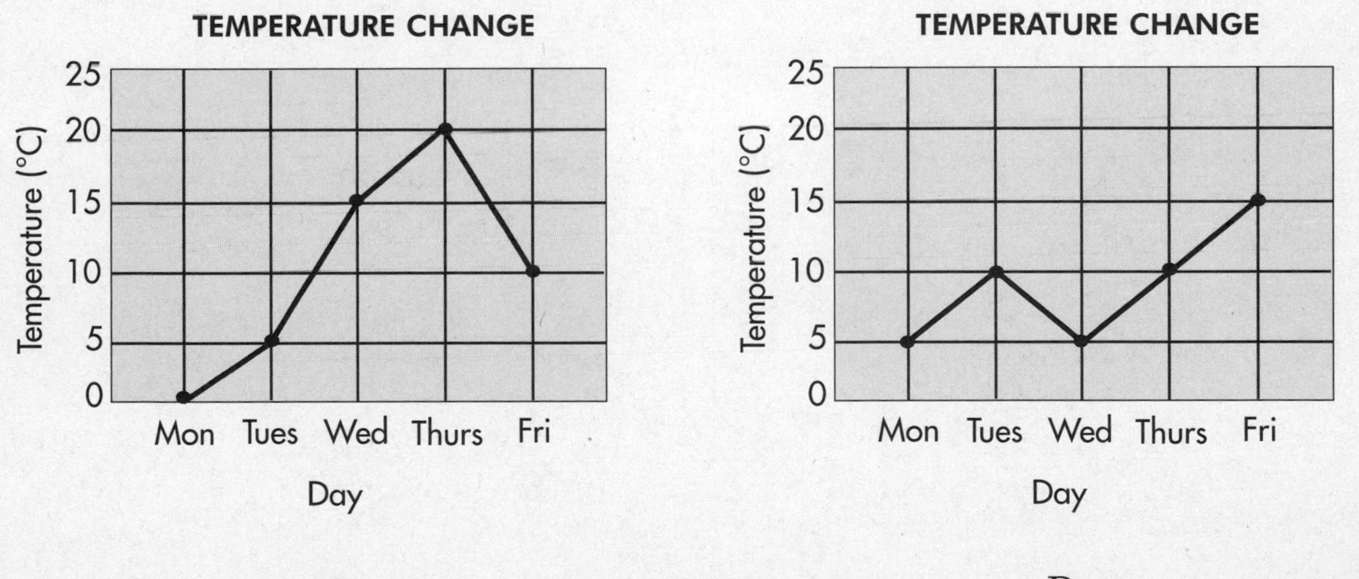

C

D

7 Write a sentence that reports your conclusion about the data.

____

____

8 What new question might this data lead you to ask?

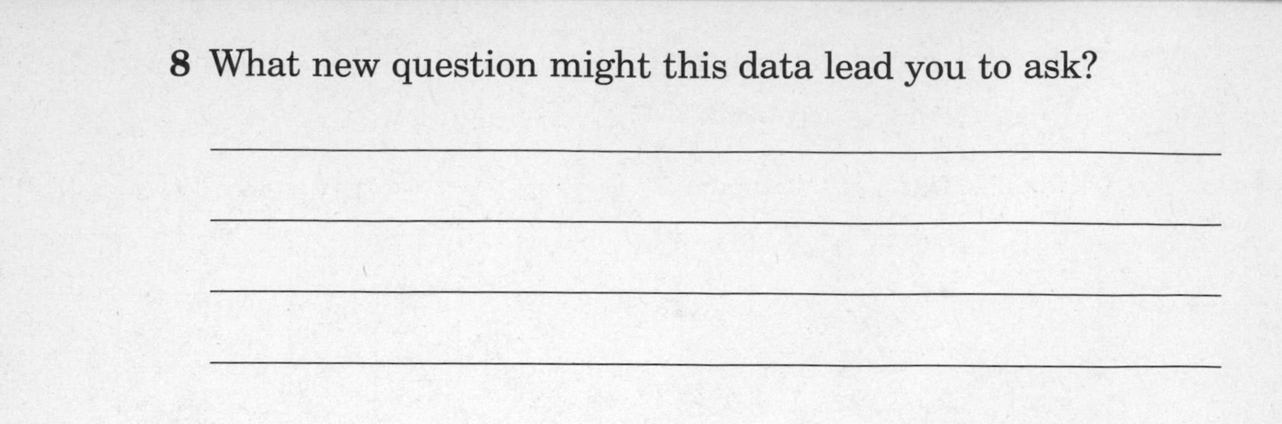

Directions (9–14): Each question is followed by four choices. Decide which choice is the best answer. Circle the letter of the answer you have chosen.

9 Which of the following would best show the changes in the height of a growing giraffe?

A chart
B graph
C paragraph
D drawing

10 Which of the following would best show the differences in the shapes of footprints of animals?

A chart
B graph
C paragraph
D drawing

11 Why do scientists put data into charts or graphs?

A It makes it easier to study the data.
B It's a good way to organize data.
C It's helpful when explaining data to others.
D All of the above.

12 Which of the following are the best headings on a data chart that shows temperatures recorded at 8 A.M. and at 8 P.M.?

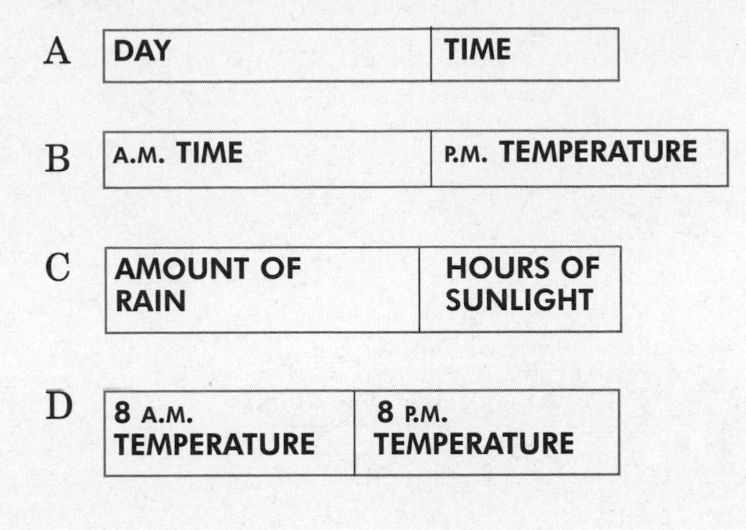

A

DAY	TIME

B

A.M. TIME	P.M. TEMPERATURE

C

AMOUNT OF RAIN	HOURS OF SUNLIGHT

D

8 A.M. TEMPERATURE	8 P.M. TEMPERATURE

---

13 Diane is testing brands of soap to see the amount of suds each can produce. Which headings should she use to record her information in a chart?

A

Weight of Soap	Cost of Soap

B

Brand of Soap	Suds Produced

C

Cost of Soap	Color of Soap

D

Color of Soap	Design of Package

14 Tom rolled a ball down each of the ramps shown below. He rolled a ball down each ramp twice. Every time he rolled the ball, he timed how many seconds it took for the ball to roll to the bottom of the ramp. He put his data into a chart. What conclusion can be made from Tom's data?

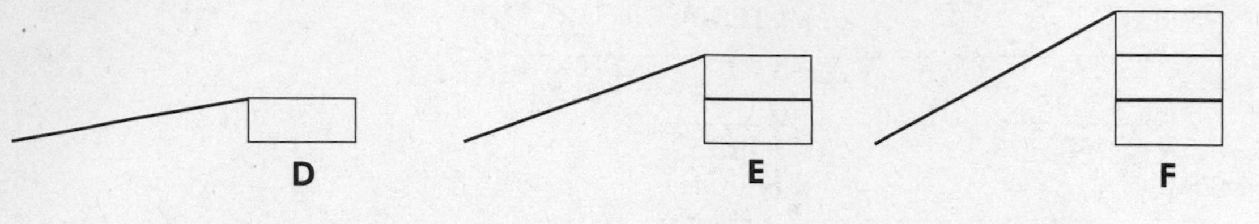

	FIRST ROLL	SECOND ROLL
Ramp D	4 seconds	5 seconds
Ramp E	3 seconds	3 seconds
Ramp F	2 seconds	2 seconds

A The ball rolled fastest down the highest ramp.
B The ball rolled fastest down the lowest ramp.
C The ball rolled fastest down the middle ramp.
D The height of the ramp did not affect how fast the ball rolled.

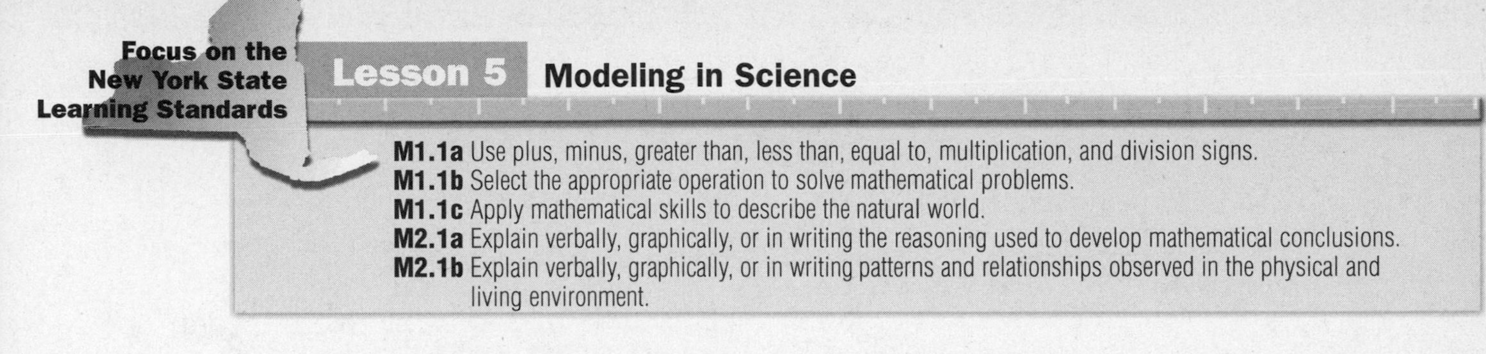

Focus on the New York State Learning Standards

Lesson 5 Modeling in Science

M1.1a Use plus, minus, greater than, less than, equal to, multiplication, and division signs.
M1.1b Select the appropriate operation to solve mathematical problems.
M1.1c Apply mathematical skills to describe the natural world.
M2.1a Explain verbally, graphically, or in writing the reasoning used to develop mathematical conclusions.
M2.1b Explain verbally, graphically, or in writing patterns and relationships observed in the physical and living environment.

You can use your math skills to help you describe the natural world.

Having good **math skills** means being able to select the correct math operation to solve a problem.

An **average** is the typical number of a whole group of numbers.

Guided Instruction

Directions Read the following information.

You begin a scientific inquiry by asking a question. You test your explanation, or the answer to your question, to see if your explanation is or is not true. You also make and record observations. Once data, or information, has been collected and organized, you often can use your **math skills** to describe your observations.

You might wish to find the **average**, or typical, number of something you have observed. For example, you might wonder how many birds live in a wooded area. It would be difficult to count each and every bird. So you might count the birds you see on several different days and put the information into a chart.

DAY OF OBSERVATION	NUMBER OF BIRDS
Monday	7
Tuesday	8
Wednesday	4
Thursday	5
Friday	6

To find the average number of birds in the woods you would use the data in your chart and do the following:

Guided Questions

Why would **math skills** be needed for science class?

- Find the total number of birds:

  $7 + 8 + 4 + 5 + 6 = 30$
- Divide the sum of 30 by the number of days, 5.

  $\frac{30}{5} = 6$

Your observation over five days is that, on average, there are 6 birds in the wooded area. You will notice that 6 birds are less than the number of birds counted on Tuesday, and 6 birds are more than the number of birds counted on Wednesday. But six is the one number typical of this whole group of numbers. Six is the average.

In science, math skills can help you find average height and average weight of animals or objects or average rainfall in a study of the climate. Finding the average is one way scientists can use math to explain patterns and relationships that they observe in the natural world.

Guided Questions

Why might a scientist want to know the average rainfall of an area?

How can math skills help you in science?

Directions For each question, write your answer in the space provided.

1. What is the first step in any scientific inquiry?

____

2. Why is it important for a scientist to have math skills?

____

3. What is an average number?

____

4. What is the average of these four numbers: 6, 4, 4, 10? Show your work.

5. How would you find the average weight of a litter of new born puppies?

____

____

Apply the New York State Learning Standards to the State Test

Directions: For each question, write your answer in the space provided. Base your answers to questions 6 through 11 on the paragraph and chart below.

Some students in your health class believe that they are quite fit. To test this, they weighed each student and then found the average weight of this group of students. Later the students will compare their average weight to the national average for their age group to find if they are more or less fit than the national average.

STUDENT	WEIGHT IN POUNDS
Lauren	110
Jack	118
Justin	113
Maria	99

6 What is the average weight of the students in this group? You may use a calculator.

____

7 How did you find the average weight?

____

____

8 Write a math sentence that shows how you found the average weight.

____

____

9 Which student weighed less than the average weight?

____

10 Which students weighed more than the average weight?

____

11 Did anyone weigh the same as the average weight? If so, who?

____

____

Directions (12–17): Each question is followed by four choices. Decide which choice is the best answer. Circle the letter of the answer you have chosen.

12 Why are math skills important in scientific inquiry?

A They help you get good grades.
B They help you keep track of scientific equipment.
C They help you describe observations.
D They help you count birds.

13 Which statement is the best description of the pattern in the following numbers?

2, 4, 6, 8, 10, 12

A Each number is 4 less than the number before it.
B Each number increases by 2.
C Each number increases by 3.
D Each number is 2 times more than the number before it.

14 Which number is the next number in the following pattern?

25, 50, 75, 100, 125, __

A 100
B 150
C 175
D 225

15 The chart below shows the temperature at noon on three days.

DAY	NOON TEMPERATURE
Monday	20°C
Tuesday	24°C
Wednesday	22°C

What was the average noontime temperature for these three days?

A 20°C
B 22°C
C 24°C
D 26°C

16 Monthly rainfall was measured in June, July, and August. There were 5 cm of rain in both June and July and 2 cm in August. Which math sentence models the average rainfall for three months?

A $\frac{12}{3} = 4$ cm
B $5 \times 2 = 10$ cm
C $5 - 2 = 3$ cm
D $12 \times 3 = 36$ cm

17 To make the seesaw balance, both sides need to be equal. What math operation should be done to balance the seesaw shown below?

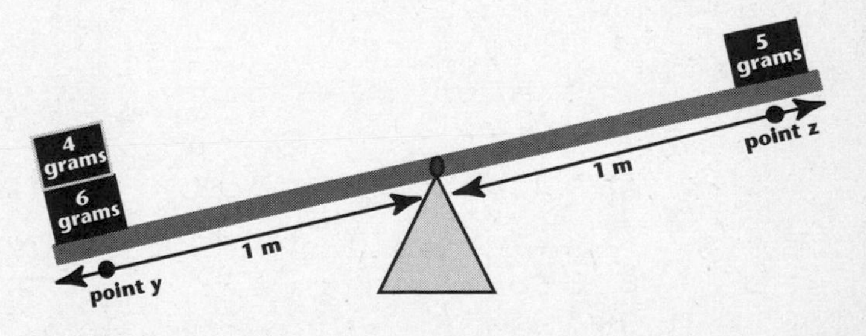

A Subtract 4 kilograms from point Y.
B Divide the weight at point Y by 5.
C Add 5 grams at point Z.
D Multiply the weight at point Z by 5.

Higher-Order Performance Task

Using Measurement Tools

Task:

You will identify the best tool to use for different measurements. You also will use the tools and record measurements. The directions below tell you how to begin.

Materials:

- metric ruler
- balance
- gram cubes
- graduated cylinder
- container of water
- 2 plastic cups
- masking tape
- paper towels
- objects to be measured:
  - —small box
  - —small rock
  - —marble

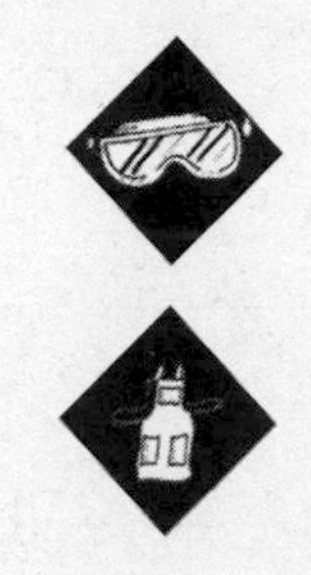

Directions:

1. Use the materials above to complete this task.

2. Observe the box. Which tool would you use to measure the volume of the box?

____

Describe how you would use this tool to find the box's volume. Remember that volume = length × width × height.

____

____

3. Measure and find the volume of the box.

Volume = ____

4. What metric unit did you use when you recorded the volume?

____

5. Observe the small rock. Which tool would you use to measure its volume?

Describe how you would use this tool to find the rock's volume.

6. Measure and record the volume of the rock.

7. What metric unit did you use when you recorded the volume?

8. Now that you know the rock's volume, how can you find the mass of the rock?

9. What metric unit will you use?

____

10. What is the mass of the rock?

____

11. Use the materials to find the mass of the marble. Describe any problems you had finding the mass of the marble.

____

____

How did you solve the problem?

____

____

____

12. When you have finished, put the materials back the way you found them. Dry any objects that are wet and wipe up any spills.

Building Stamina®

Directions (1–18): Each question is followed by four choices. Decide which choice is the best answer. Circle the letter of the answer you have chosen.

1 Where would you be if you recorded the following observations: loud noise, winged objects, suitcases, runway, and many people?

A baseball game
B airport
C train station
D expressway

2 Earth orbits once around the Sun about every 365 days. The Moon orbits around Earth. About how many days does it take the Moon to orbit twice around the Sun?

A 730
B 365
C 64
D 29

3 The scent is strong and your eyes fill with tears and they sting terribly. Which question would a scientist ask?

A Am I sad?
B Is there a pollutant in the air?
C What's for dinner?
D Am I feeling cold?

4 You observe the following features of an animal: horns, gives milk, lives on a farm, and eats almost anything. What can you conclude about the animal?

A It is a sheep.
B It is a cow.
C It is a steer.
D It is a goat.

Base your answers to questions 5 and 6 on the graph below.

FRUIT

NUMBER OF VOTES	APPLES	PEACHES	BANANAS
6			
5			
4			
3			
2			
1			

5 Which fact is represented on the graph?

A Bananas received twice as many votes as peaches.
B Peaches received twice as many votes as apples.
C Peaches received twice as many votes as bananas.
D Bananas received all of the votes.

6 What can you infer from the data on the graph?

A There are only three kinds of fruits.
B Bananas are liked the best.
C Peaches are liked by the majority of the voters.
D Bananas are liked by 12 of the voters.

7 Your experiment found that a fossil was 1,000 years old. Another study found it to be 100,000 years old, and a third study found it to be 110,000 years old. How can you explain your answer?

A The other two studies are wrong.
B Your findings may be wrong.
C The fossil's true age is between 1,000 and 110,000 years old.
D The fossil is older than 110,000 years.

8 An amoeba can divide itself into two separate animals. If you start with 1 amoeba and end up with 4, how many divisions occurred?

A 8
B 4
C 3
D 2

9 Look at the chart below. How could you change it if you wanted to grow the plants for another week?

	PLANT 1	PLANT 2
Height at 1 week	1 cm	1 cm
Height at 2 weeks	5 cm	3 cm

A Add a column for observations.
B Add a column for average height.
C Add a row for soil conditions.
D Add a row for height at 3 weeks.

10 After setting your books down on the lab desk, you notice that the desk and now your books are wet. What should you do?

A Tell your teacher and help her wipe it up.
B Wipe it up quickly, before anyone else puts their books in the liquid.
C Ignore it. It is probably just water.
D Look around to see if other lab desks are wet.

11 Which is not an observed fact?

A The Sun rises in the east.
B Flowers bloom in spring.
C It always rains on Mondays.
D It usually gets cooler at night.

12 The copier cut off the last sentence of the lab instructions. What should you do?

A Ask your teacher what it says before continuing.
B Guess about the last step based on the information in the previous steps.
C Begin your experiment and ask your teacher when you get to that point.
D Skip that step. The last one can't be too important anyway.

13 Which equipment goes together?

A centimeter ruler, milliliter beaker
B inch ruler, milliliter beaker
C milliliter beaker, measuring cup in ounces
D a centigram weight, measuring spoon in ounces

14 Julie presented the results of her experiment to the class and answered their questions. What purpose did this serve?

A She wanted to show off how much she knows.
B She wanted feedback and comments on her results.
C She wanted to check to see if anyone else had already done this experiment.
D She wanted to get a good grade.

15 A geologist is counting fossils in a piece of limestone. Her data is recorded in the circle graph below. Which summary is correct?

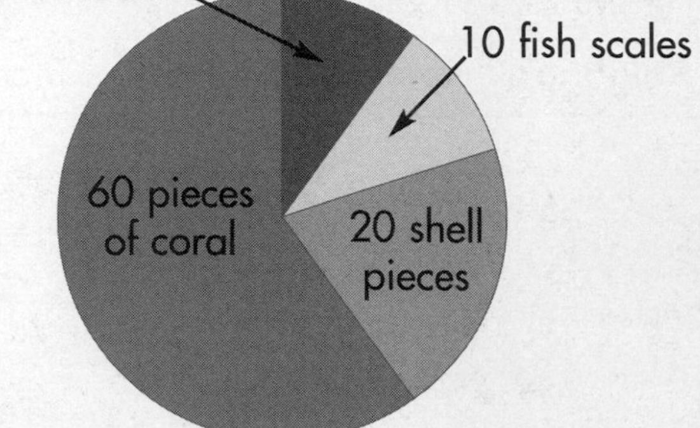

A Most of the fossils are not coral.
B Most of the fossils are trilobites.
C Most of the fossils are coral.
D Half of the fossils are fish scales.

16 The beaker you are about to use has a crack in it. What should you do?

A Continue with your experiment. It is only a small crack.
B Throw the beaker away and take another.
C Show another student the crack and continue with your work.
D Show your teacher and wait for instructions.

17 Safety goggles should be worn during all but which part of an experiment?

A set up
B running the experiment
C clean up
D leaving the room

18 You have begun a timed experiment and suddenly realize that you are missing a thermometer for the next step. What should you do?

A Quickly walk away to look for a thermometer, and try to get back in time.
B Borrow the thermometer from the group next to you.
C Tell your partner to watch the experiment until you get back.
D If possible, stop the experiment, get the thermometer, and begin again.

Directions (19–27): For each question, write your answer in the space provided.

Base your answers to questions 19 through 21 on the graph below.

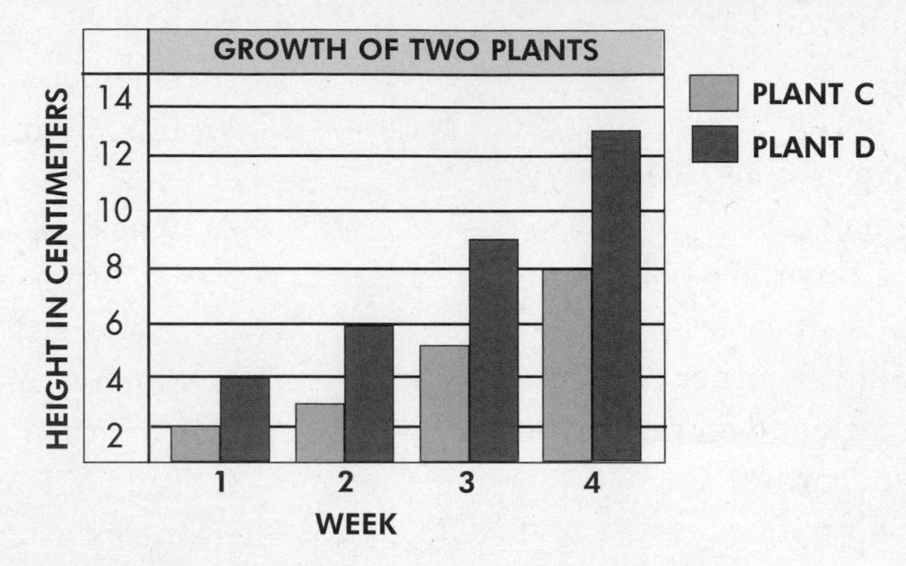

19 Look at the graph. What information does this graph show?

____

____

20 What is another way the data could be presented?

____

21 What sort of generalizations and conclusions can you make about the plant data?

____

22 Arrange the letters of the steps for this experiment in the correct order:

a. perform investigation
b. make conclusions
c. record observations
d. gather needed materials
e. read instructions
f. observe
g. clean up
h. wait for your teacher's instructions

____

23 If you are wearing old clothes and don't care if they get stained in science class, why wear an apron?

____

24 Explain and show the steps you would use to find the average of the following set of numbers: 18, 7, 12, and 3.

What is the average? ______________________________

25 Explain how you could use words and graphics to present your leaf collection to a visitor.

Words: ______________________________________

____

Graphics: ____________________________________

____

26 Explain how you might test your explanation that dogs will eat pickles.

____

____

27 You need to determine and record the mass of a certain amount of liquid. What equipment do you need?

____

____

Focus on the New York State Learning Standards

Lesson 6 Characteristics and Needs of Living and Nonliving Things

1.1a Animals need air, water, and food in order to live and thrive.
1.1b Plants require air, water, nutrients, and light in order to live and thrive.
1.1c Nonliving things do not live and thrive.
1.1d Nonliving things can be human-created or naturally occurring.
1.2a Living things grow, take in nutrients, breathe, reproduce, eliminate waste, and die.
5.1a All living things grow, take in nutrients, breathe, reproduce, and eliminate waste.

Living things are both similar to and different from each other and from nonliving things.

Nutrients are substances a living thing needs for energy and growth.

Reproduction means producing young, or more of your own kind.

Guided Instruction

Directions Read the following information.

The world around us is made up of both living and nonliving things. A plant growing in a flowerpot is living, but the flowerpot is not alive, so it is called a nonliving thing. Scientists use characteristics to group or classify something as a living thing. For example, all living things grow during their lives. Therefore, something that grows is classified as a living thing.

In order to live and be healthy, all animals need air, water, and **nutrients** in food. Animals use the nutrients for energy to grow and stay healthy.

Guided Questions

Where do animals get **nutrients?**

Name the living and nonliving things in the picture.

Plants need air, water, and nutrients, too. They use sunlight to make their own food, using nutrients from the water and soil. Therefore, plants need air, water, nutrients, and light in order to live.

Animals cannot make their own food. For energy and growth, animals must eat. As they use the nutrients they take in, animals release waste products. Animals also release waste products when they breathe. Animals release carbon dioxide as a waste product.

Most plants can make their own food using water, air, and sunlight. The nutrients plants take in from the soil also help them grow strong and healthy. Plants release waste products too. They release oxygen as a waste product.

All living things can make more of their own kind through **reproduction**. Sunflowers produce seeds that grow into new sunflowers. Cats have litters of kittens, dogs produce puppies, and human beings have babies.

The last characteristic of living things is that at the end of their life cycles, they die. Some plants live only one year, and others live for many years. All animals die at the end of their life spans.

Nonliving things are not alive, so they do not need air, water, or food. Nonliving things cannot reproduce. Many nonliving things are part of the natural world, such as rocks. Other nonliving things have been made by human beings. Cars, for example, are made by factory workers, using nonliving substances such as metal and plastic. Even though cars need gasoline, water, and air to move, they are nonliving.

Guided Questions

Where do plants get **nutrients?**

What is **reproduction?**

Where do nonliving things come from?

Directions For each question, write your answer in the space provided.

1. What do all living things need in order to live and stay healthy?

2. What other things do plants need—that animals do not need—in order to live? Why?

____

____

3. What do all living things do?

____

____

4. An icicle "grows" as more water freezes. Is an icicle a living thing? How do you know?

____

____

5. Explain why cats cannot produce puppies.

____

____

6. In order to run, a car needs air, water, and its own kind of "food." It also eliminates wastes and sometimes a car can "die." How is a car the same as a living thing? How is it different from a living thing?

____

____

____

____

Apply the New York State Learning Standards to the State Test

Directions: Mark an "X" in the appropriate rows below to answer items 7 through 9. Then answer question 10.

	ARE FOUND IN NATURE	ARE MADE BY HUMAN BEINGS	NEED AIR, WATER, AND NUTRIENTS	REPRODUCES ITS OWN KIND	ELIMINATES WASTES
7 Living Things					
8 Nonliving Things					
9 Both Living Things and Nonliving Things					

10 Suppose you put a hamster and a plant in a very dark room. You give them both water every day. You also feed the hamster and give the plant nutrients. Explain what will happen to the hamster and the plant.

____

____

Directions (11–16): Each question is followed by four choices. Decide which choice is the best answer. Circle the letter of the answer you have chosen.

11 What do all animals need in order to live?

A air, water, and light
B air, water, and nutrients
C air and water
D air, nutrients, and light

12 Which of the following things is living?

A rock
B sand
C air
D seaweed

13 Which of the following things needs light to make food?

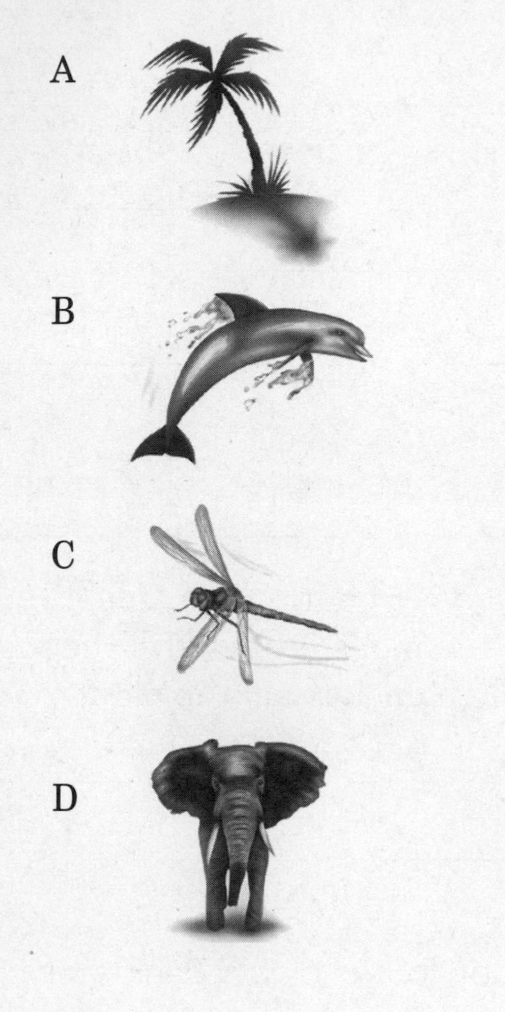

14 Which of the following things can grow bigger but is nonliving?

A grass
B beetle
C icicle
D mushroom

15 What does a car do that a living thing also does?

A It uses air and water.
B It needs light, water, and nutrients.
C It can move from place to place by itself.
D It has adapted to its environment.

16 Suppose you see some dark specks floating in a jar of water. You put a lid on the jar and look at it the next day. What might make you think the specks are tiny living things?

A They have settled to the bottom.
B They are still floating in the water.
C The number of specks has doubled.
D The specs are gone and the water is gray.

Focus on the New York State Learning Standards

Lesson 7 Traits of Living Things

2.1a Some traits of living things have been inherited.
2.1b Some characteristics result from an individual's interactions with the environment and cannot be inherited by the next generation.
2.2a Plants and animals closely resemble their parents and other individuals in their species.
2.2b Plants and animals can transfer specific traits to their offspring when they reproduce.

Traits of living things can be inherited, acquired, or learned.

Traits are qualities or characteristics of a living thing or a species.

Inherited traits are passed down from parents to offspring.

A **species** is one of the groups into which scientists divide living things according to their shared characteristics. Living things reproduce members of their own species.

Offspring are new living things that parents produce, or the young of plants and animals.

Guided Instruction

Directions Read the following information.

All living things have characteristics that are special. Most fish have one eye on each side of its head. Bean plants have green leaves, and birds have two wings. Characteristics of living things are called **traits**.

Some traits can be **inherited** and some can be learned. Inherited traits are passed from parents to their young. For example, frogs are able to swim when they are born. A parent frog will pass on the ability to swim to its young. The ability to swim is a trait that must be learned by humans. It is not an inherited trait. Even if a mother and father are champion swimmers, their sons or daughters can swim only if they are taught.

Guided Questions

Give an example of a **trait.**

The dog waits to eat. Is this an **inherited** trait or a learned trait?

A living thing can develop a new characteristic after it is born. These characteristics cannot be inherited or passed on. For example, you can build large muscles by lifting weights. You can dye your hair blond. A parrot can learn to say human words. But these traits cannot be passed on to offspring.

Most living things look very much like other members of their **species**. No two tigers have stripes in exactly the same place, but you can tell at a glance that each one is a tiger. Tigers belong to the same species. A species is a group of living things that share characteristics and can produce offspring. All human beings belong to the same species. All dogs belong to the same species too.

Plants, animals, and human beings can pass on some special traits to their own **offspring** when they reproduce. For example, if two black dogs have puppies, most of their puppies will probably be black. If two large cats have kittens, the kittens will probably grow up to be large. Living things usually look even more like their parents than like other members of their species. You have probably noticed that many people you know look very much like their family members.

Guided Questions

What is a **species?**

What are **offspring?**

Directions For each question, write your answer in the space provided.

1. Give an example of a trait that you inherited.

____

____

2. Give an example of a characteristic you learned.

____

____

3. Explain if you can inherit a suntan.

____

____

4. Why is the ability to swim an inherited trait in a fish but a learned trait in a human being?

____

____

____

5. If a dog's owner cut its long fur very short, would the dog's puppies have very short fur, too? Explain.

____

____

____

6. Choose a living thing. Explain what traits all members of that kind of living thing share. Then explain what traits might be different in offspring born to different parents.

____

____

____

Apply the New York State Learning Standards to the State Test

Directions: For each question, write your answer in the space provided. Use the picture to answer questions 7 through 11.

7 List one way all of the cats in the picture are alike.

____

8 List one way the cats in the picture are different.

____

9 List three traits that are inherited by *all* members of the cat species.

____

____

____

10 List two traits that are inherited in only one cat.

____

____

11 The white cat lost part of its ear in an accident. Explain why this will not be passed on to her offspring.

____

____

Directions (12–17): Each question is followed by four choices. Decide which choice is the best answer. Circle the letter of the answer you have chosen.

12 Which trait is inherited?

A language in a human being
B speech in a parrot
C feathers on a duck
D a brand on a steer

13 Which trait is *not* inherited by a human being?

A short hair
B brown hair
C blue eyes
D brown eyes

14 Which trait is inherited by all members of the same species?

A black fur on a dog
B two wings on a robin
C gray eyes on a human being
D two flowers on a rose bush

15 Which trait is inherited by only some members of a species?

A long neck on a giraffe
B green leaves on an elm tree
C sharp quills on a porcupine
D black legs on a horse

16 Which trait can offspring inherit from their parents?

A skill in writing
B ability to think
C ability to juggle
D skill in basketball

17 Which trait will puppies *not* inherit from their parents?

A speed
B strength
C ability to obey commands
D ability to bark and growl

Focus on the New York State Learning Standards

Lesson 8 Characteristics of Animals

3.1a Each animal has different structures that serve different functions in growth, survival, and reproduction.
5.1b An organism's external physical features can enable it to carry out life functions in its particular environment.

Throughout time, animals have changed and adapted to their environment.

Living things **adapt**, or go through changes that make them fit in better with the environment around them.

An **organism** is a living thing, such as a plant or animal.

A **predator** kills and eats animals.

Guided Instruction

Directions Read the following information.

Living things are called **organisms**. All animals and plants are organisms. One characteristic of organisms is that they **adapt** to their environment. Adapt means that over time a species will change their structures and behavior to help them survive in their environment.

Animal species have developed different structures to do different jobs over time. Some structures help animals grow, others help them reproduce, and still others help them survive. For example, let's say the color and markings of a cheetah cub's fur match its grassy environment. This helps the baby cheetah survive because a **predator** will not see it. Then the cheetah will grow and reproduce more cheetah cubs with the same kind of fur. After a while all the cheetahs will have the same kind of fur. Another example is the

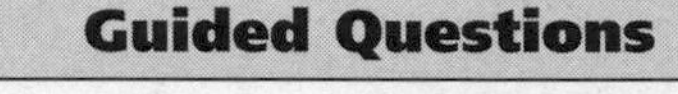

Guided Questions

What are **organisms?**

What does **adapt** mean?

What are **predators?**

How have organisms in the picture adapted to their environments?

neck of a giraffe. The giraffes with longer necks could reach the tallest trees, so they could eat. The giraffes with longer necks lived and reproduced more giraffes with long necks. Over time, all the giraffes had long necks.

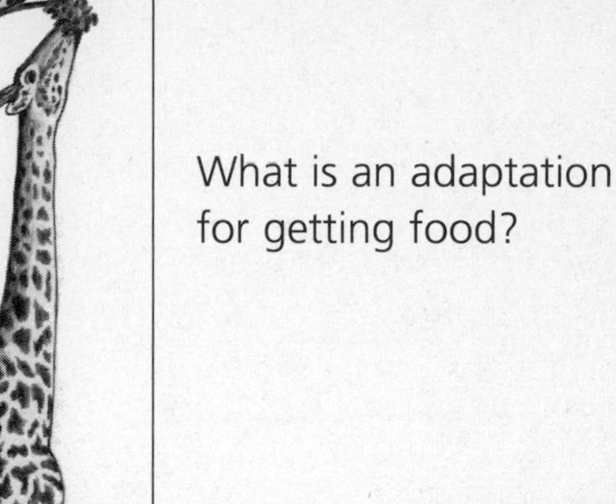

Wings, legs, or fins help some animals escape from predators. Other animals developed sharp senses, such as eyesight, smell, taste, and touch for protection. Some animals, such as the skunk, developed a strong scent to keep animals away. Still others developed sharp teeth and claws to help them protect themselves. The sharp teeth also help animals hunt food. Once animals find food, they need to be able to eat it. Many animals have mouths, teeth, jaws, and tongues to help them eat and drink.

Other species have developed body coverings, such as shells and spines that protect them from predators. The color of these body coverings may help animals blend into their surroundings and hide from predators. Coverings such as feathers, fur, and scales also help protect against heat, cold, wetness, dryness, and other conditions in the environment.

Another characteristic of living organisms is that they try to keep their bodies at the right temperature. Some animals grow longer fur in winter to keep warm. Other animals build up body fat to stay warm during the winter. Some animals hibernate and others fly south for the winter. If the environment begins to change, the characteristics and traits of species may adapt over time again to help animal species survive.

Guided Questions

What is an adaptation for getting food?

What is an adaptation for protection?

What is an adaptation for dealing with cold weather?

Directions For each question, write your answer in the space provided.

1. Why did species of organisms change over time?

____

____

2. What are three traits that help animals protect themselves from predators?

____

____

____

3. How might making a scent protect an animal from predators?

____

____

4. Why do you think some small brownish animals, such as arctic foxes, have adapted by turning white during the winter?

____

____

Apply the New York State Learning Standards to the State Test

Directions: For each question, write your answer in the space provided. Use the pictures to answer questions 5 through 7. Then answer questions 8 through 9.

5 What structures might help each animal above escape predators?

bird: _______________

snake: _______________

6 What structures might help each animal on page 52 find food?

bird: ______________________________

snake: ______________________________

7 How can you tell if a bird is a hunter of other animals?

____

8 Why is it important for a species to avoid predators?

____

____

9 What could happen to a species of animals if they did not adapt to their environment?

____

10 What would happen to polar bears at the North Pole if the environment suddenly became very warm?

____

____

____

Directions (11–16): Each question is followed by four choices. Decide which choice is the best answer. Circle the letter of the answer you have chosen.

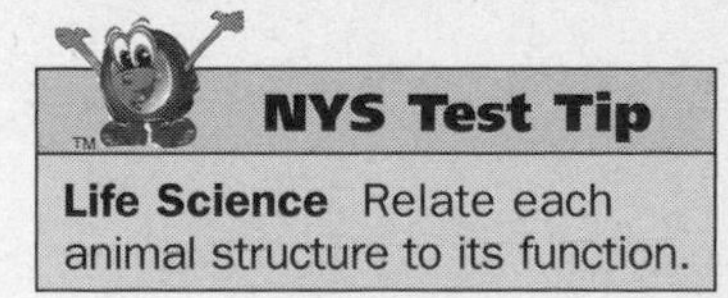

11 Which structures can help animals get information about their surroundings?

A eyes, noses, ears, fur, and skin
B eyes, noses, ears, teeth, and skin
C eyes, noses, ears, tongue, and skin
D eyes, noses, ears, tails, and skin

12 Which structures do many animals have to help them eat and drink?

A mouths, teeth, jaws, noses, and tongues
B mouths, teeth, jaws, and tongues
C mouths, teeth, jaws, feet, and tongues
D mouths, teeth, jaws, and lips

13 When would an animal be most likely to grow longer hair?

A winter
B spring
C summer
D fall

14 Which animal is best adapted to quickly flee its predators?

A raccoon
B elephant
C deer
D chicken

15 Which bird would be most likely to use its wings to hunt other animals?

16 What would be the safest color for leaf-eating insects?

A brown
B black
C red
D green

Focus on the New York State Learning Standards

Lesson 9 Characteristics of Plants

3.1b Each plant has different structures that serve different functions in growth, survival, and reproduction.
5.1b An organism's external physical features can enable it to carry out life functions in its particular environment.

Throughout time, plants have changed and adapted to their environment.

Energy is what gives the plant strength to live, grow, and carry out life processes. When a seed **germinates**, it starts to grow roots and shoots.

Directions Read the following information.

All organisms need **energy**. Just as a car needs gas to make it run, organisms need food or some other form of energy to live and thrive.

You may have noticed that almost all plants have green parts, but very few animals do. Plants need some structures, like leaves, that animals do not need. You may remember that plants can make their own food. They use the **energy** in sunlight to make food. Leaves help plants gather sunlight and use it to make food for the plant. Most leaves are green.

Besides sunlight, a plant needs water and nutrients in order to make food and use energy. Roots grow in the soil and take in water and nutrients. Carrots, beets, and turnips are all roots with stored energy and nutrients. Roots also help support a plant. Plants have other structures that help support them too. For example, dandelions have stems, celery has stalks, and pine trees have trunks.

You have seen flowers on many plants. Flowers help the plant reproduce. They produce fruit, and

Guided Questions

What does **energy** do for a plant?

What are two functions of roots?

the fruit contains seeds. In the spring, before a plum tree has fruit, it is covered with flowers. The plums that grow from these flowers have seeds inside. Most flowering plants grow from seeds. Seeds contain stored food that helps the seeds **germinate**. They begin to grow roots and shoots. This helps the young plants grow. Because of this stored food, seeds such as beans and nuts make healthful food for animals and humans.

Like other organisms, plants adapt to their environment. Their structures change over time to help them survive better in their environment. For example, a cactus has a thick stem that can store water. The structures of plants in dry places are different from those of plants in wet places. The structures of plants in cold or warm places differ too. The environment of the plant species affects how deep the roots are, how thick or broad the leaves are, and how tall and strong the stems and trunks are.

Guided Questions

What does **germinate** mean?

Directions For each question, write your answer in the space provided.

1. What structures help support plants?

___

2. What structure of a carrot plant stores food?

___

3. What structures do plants need that animals do not need? Why?

___

4. What would happen if an apple tree did not have any flowers one year?

___

5. Where do seeds get energy to germinate?

____

6. Why would plants need different structures to survive in dry climates?

____

____

7. Why are beans and nuts healthful foods?

____

Apply the New York State Learning Standards to the State Test

Directions: For each question, write your answer in the space provided. Use the pictures to answer questions 8 through 13.

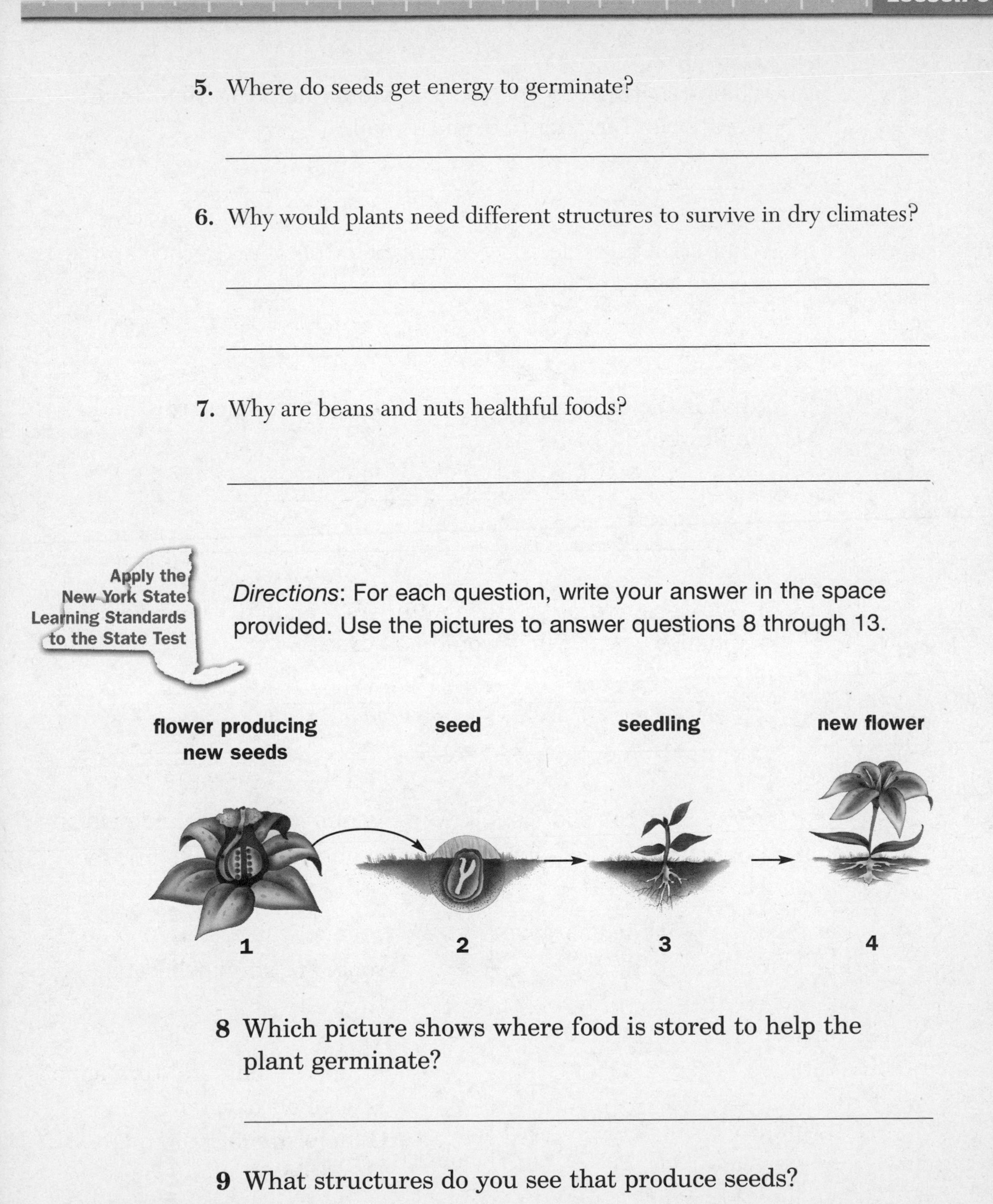

8 Which picture shows where food is stored to help the plant germinate?

____

9 What structures do you see that produce seeds?

____

10 What structures do you see that would help the plant take in water and nutrients from the soil?

____

11 What structures do you see that help the plant gather and use the Sun's energy?

____

12 What is the difference between the flower in picture 1 and the flower in picture 4?

____

____

13 If you were to draw picture numbers 5 and 6 to show what will happen next, what would you draw?

____

____

Directions (14–19): Each question is followed by four choices. Decide which choice is the best answer. Circle the letter of the answer you have chosen.

14 What gives a plant the energy it needs to make food?

A nutrients
B water
C roots
D sunlight

15 Stems, stalks, trunks, and roots are similar because they all

A have a long and thin shape
B give off oxygen
C help support the plant
D gather sunlight for the plant

16 What is happening to the seed below?

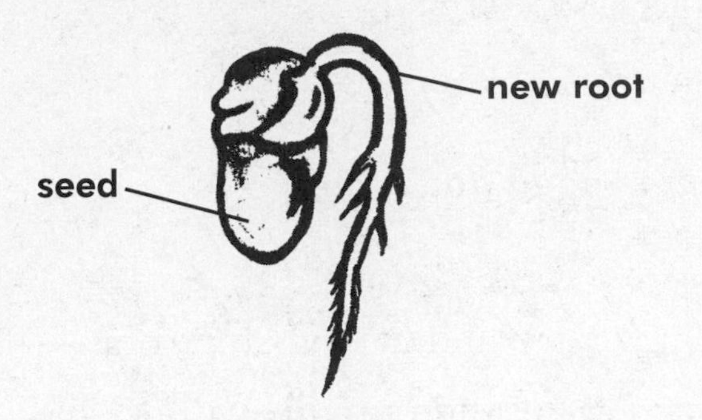

A It is storing food for the young plant.
B It is germinating.
C It is dying.
D The seed is coming out from inside the fruit.

17 Why do you think some plants make good tasting fruit?

A to get animals to eat the fruit and drop the seeds
B to help animals and humans survive
C to supply fruit for stores
D to be sure the seed inside has plenty of stored food to germinate

18 Why do you think roots can be healthful foods?

A They gather energy from the Sun.
B A plant cannot grow if separated from its roots.
C They have more water than other parts of the plant.
D They take in nutrients and sometimes store energy.

19 What might happen to a plant if you pulled off all its leaves?

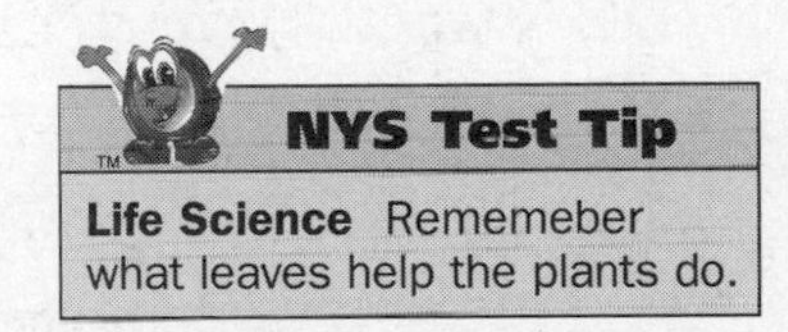

A It would produce more flowers, because it would have energy to spare.
B It would grow new leaves, because it needs them to gather sunlight.
C It would die, because it could not get food and energy for growth.
D It would use its stems to gather energy from the Sun.

Focus on the New York State Learning Standards

Lesson 10 Adaptations

3.1c In order to survive in their environment, plants and animals must be adapted to that environment.
3.2a Individuals within a species may compete with each other for food, mates, space, water, and shelter in their environment.
3.2b All individuals have variations, and because of these variations individuals of a species may have an advantage in surviving and reproducing.

Throughout time, organisms have changed and adapted to their environment.

To **hibernate** is to go to sleep for the winter and live on stored fat.

To **migrate** is to move from one place to another, usually with the change of seasons.

Guided Instruction

Directions Read the following information.

In order to survive in their environment, plant and animal species must be adapted to that environment. Plants in different environments have different leaves, flowers, stems, and roots. These structures may be different in size, shape, thickness, color, and scent. Structures of living things are different to fit the environment and the needs of the species. For example, plants of the desert, such as cactus, store water in their leaves and trunks. They have small needle-like leaves so water doesn't easily evaporate. Many desert plants store the Sun's energy but don't make food during the hot daytime, so that they do not lose water.

Guided Questions

What are some adaptations of a cactus?

Seeds need space, light, nutrients, and water in order to grow. So parent plants need to spread their seeds far away from themselves. Species of plants have also adapted ways to spread their seeds. Plants that depend on wind to carry seeds have seeds that are tiny and light or have wing-like structures. Plants that live near moving water may have seeds or fruit that float. Some plants depend on animals to spread their seeds. These plants must make tasty, colorful fruit to attract animals.

Animal species have adapted their behaviors to survive seasonal changes. Some animals may **migrate** to warmer or cooler climates. You may have noticed that you see certain birds only in the spring and summer. Perhaps you have seen geese or crows flying north or south. Other animals, such as chipmunks and woodchucks, **hibernate** during the winter by living on stored fat.

In nature, organisms of a species compete fiercely for food, space, light, water, and mates. Individual differences give some members of a species a better chance of surviving and reproducing. For example, a tall tree gets more Sun than the smaller trees that live in its shade. The peacock with the brightest tail has the best chance of attracting mates and reproducing.

Guided Questions

How does **migration** help a species survive in the environment?

What does **hibernate** mean?

Directions For each question, write your answer in the space provided.

1. How is a cactus adapted to its environment?

____

____

____

____

2. What is special about seeds that need wind to carry them?

____

____

3. What are two behaviors animals have to protect themselves from cold winters?

1. ______________________________

2. ______________________________

4. What do you think animals must do before they hibernate?

____

____

5. From what environments would animals be most likely to migrate?

____

Apply the New York State Learning Standards to the State Test

Directions: For each question, write your answer in the space provided. Use the pictures to answer questions 6 through 9.

6 What might be two reasons that a cactus has spiny leaves?

____

____

Cactus

7 A pill bug can roll itself into a ball when it is frightened. How might this adaptation help it survive?

____

____

____

Pill Bug

8 Explain why an elephant may have better hearing than eyesight.

Elephant

____

____

9 What would happen to an elephant if it lost its trunk or was born without one?

____

____

Directions (10–15): Each question is followed by four choices. Decide which choice is the best answer. Circle the letter of the answer you have chosen.

10 The term for when an animal sleeps for the winter is

A migrating
B hibernating
C hiding from winter predators
D growing extra warm fur

11 What word means to move to another place for the winter?

A migrate
B hibernate
C mutate
D evacuate

12 What helps the dandelion spread its seeds?

A wind
B water
C animals
D forest fires

13 How does having the brightest tail help a peacock?

A It helps it blend in with its surroundings.
B It frightens predators away.
C It helps it gather energy from the Sun.
D It helps it attract mates and reproduce.

14 Besides having fruit that tastes good, what other adaptation might help a plant that depends on animals to spread its seeds?

A tiny seeds
B seeds that stick to fur
C seeds with wing-like structures
D seeds that float

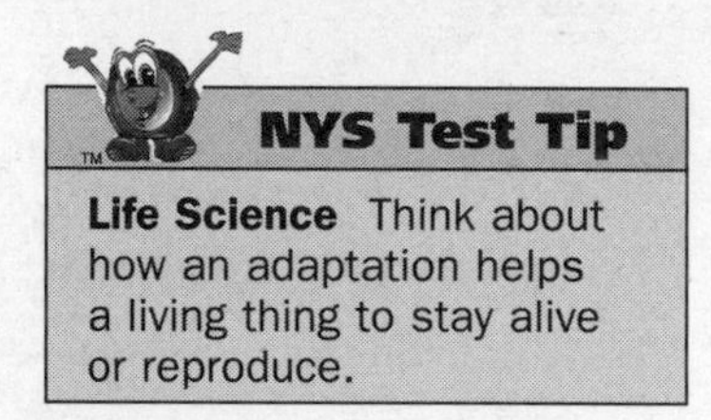

NYS Test Tip

Life Science Think about how an adaptation helps a living thing to stay alive or reproduce.

15 Animals that wake up after hibernating will probably be very

A tired
B fat
C hungry
D warm

Focus on the New York State Learning Standards

Lesson 11 Response and Behavior

5.2a Plants respond to changes in their environment.
5.2b Animals respond to change in their environment (e.g., perspiration, heart rate, breathing rate, eye blinking, shivering, and salivating).
5.2c Senses can provide essential information (regarding danger, food, mates, etc.) to animals about their environment.
5.2d Some animals, including humans, move from place to place to meet their needs.
5.2e Particular animal characteristics are influenced by changing environmental conditions including: fat storage in winter, coat thickness in winter, camouflage, shedding of fur.
5.2f Some animal behaviors are influenced by environmental conditions. These behaviors may include: nest building, hibernating, hunting, migrating, and communicating.

Organisms respond to changes in their environment.

Behavior is the way an organism responds to a change in the environment.

To **perspire** is to release extra heat by letting water escape through the skin.

Guided Instruction

Directions Read the following information.

When environmental conditions change, living things respond or also change. For example, the leaves of some green plants change position as the direction of light changes. Parts of some plants change with the seasons. Fruit and seeds leave the plant, leaves may change color and drop. Later new leaves and flowers grow.

Animals also respond to changes in the environment. When the weather warms, they **perspire**. When it cools, they shiver. Other changes cause their eyes to blink, or speed up their hearts and breathing. Animals learn about environmental changes through their senses. This information can warn of danger or help find food and mates.

Guided Questions

How do plants respond to changes in the environment?

What do animals do when they **perspire?**

Why is this tree losing its leaves?

Other animal **behaviors** are influenced by environmental conditions. Birds and other animals build nests when the seasons and the conditions are right for the eggs and the young. Many animals survive cold months by hibernating. Some animals migrate or move from place to place to meet their needs. During cold seasons, they may migrate to a warmer climate where they can find food and heat.

Humans also move from place to place to meet their needs. Since early times, humans have followed migrating game from season to season. In modern times, some people have winter homes in warm climates and summer homes in cooler climates. Farm workers may follow the crops from place to place.

Besides behavior, certain animal characteristics are influenced by changing environmental conditions. For example, animals may store fat or grow thick coats to prepare for winter. They might also change fur color to white for camouflage in the snow. When the weather warms, they shed their winter fur and fat, and change color again.

Guided Questions

What animal **behavior** is shown by these two bears?

Directions For each question, write your answer in the space provided.

1. Give two examples of how a plant responds to changes in the seasons.

____

____

2. How does perspiring make you feel cooler?

____

____

3. What would you expect a plant to do if you moved it into a room with only one window?

____

____

4. Why don't polar bears live in New York?

____

____

____

5. Compare the reasons animals move from place to place with the reasons why humans today may move from place to place.

____

____

____

____

Apply the New York State Learning Standards to the State Test

Directions: For each question, write your answer in the space provided. Use the chart below to answer questions 6 through 11. Write your answers on the write-on lines below each question.

ENVIRONMENTAL CHANGE	POSSIBLE ANIMAL RESPONSE
a. Cold Season Arrives	
b. Warm Season Approaches	
c. Favorite Food Source Disappears	
d. Sudden Danger	
e. Hot Day	
f. Cold Day	

6 In which row would you place *shivering?*

____

7 In which row would you place *perspiring?*

____

8 Suggest two possible responses for Row c: Favorite Food Source Disappears.

____

____

9 Suggest two possible responses for Row d: Sudden Danger.

____

____

10 Suggest at least two responses for Row a: Cold Season Arrives.

____

____

11 Suggest at least two responses for Row b: Warm Season Approaches.

____

____

Directions (12–17): Each question is followed by four choices. Decide which choice is the best answer. Circle the letter of the answer you have chosen.

12 Plants change with the seasons. Which is **not** a seasonal change?

A Fruits and seeds fall off the plant.
B Leaves change color.
C Flowers close their petals in the dark.
D New flowers begin to grow.

13 Why might this animal's fur color change to white in winter?

A to prepare for hibernation
B to blend in with the white snow
C to migrate more easily
D to gain a thicker coat

14 Which might be a reason for animals to move from one place to another place?

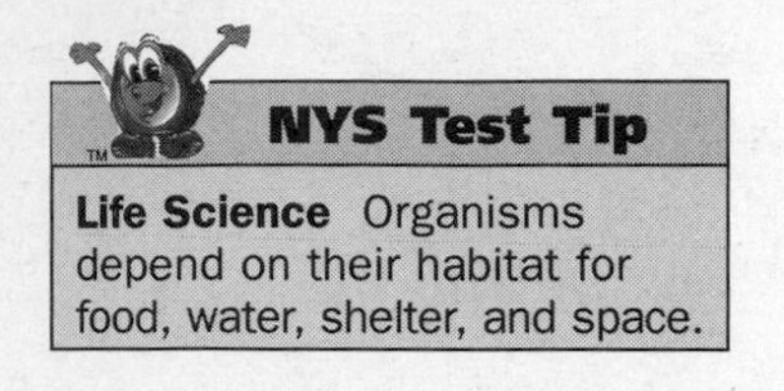

A food
B comfortable temperatures
C safe place to reproduce
D all of the above

15 How would an animal respond to a sudden bright light?

A perspiring skin
B shivering
C blinking eye
D rapid heartbeat

16 Wolves usually hunt rabbits and deer. If you saw a wolf hunting mice, what would be the most likely reason?

A Mice are less dangerous than larger animals.
B The wolf found that mice taste better than larger animals.
C There are lots of mice.
D There are no rabbits and deer.

17 When would you expect an animal to start storing fat?

A in the middle of winter
B when food starts to become hard to find
C when there is lots of food
D in early spring

Higher-Order Performance Task

Grouping Living Things

Task:

You will be putting pictures of living things into groups. The directions below tell you how to group the living things.

Materials:

- 10 pictures of living things

—fish	—worm
—gerbil	—butterfly
—bird	—pine tree
—maple tree	—rose bush
—cactus	—grass

Directions:

1. Take the pictures out and check to see that you have all of the ten pictures listed above.

2. Observe each picture of a living thing. As you study the traits of the living things, place all the animals in one pile and all the plants in another pile.

3. Write the names of the living things you classified as plants in the chart on page PT73 under the heading "Plants."

4. Write the names of the living things you classified as animals in the chart on page PT73 under the heading "Animals."

5. Look at the pictures of the plants. Notice that the plants have different kinds of leaves. The structure of some leaves helps keep the plants from losing water in a dry environment. These leaves are like sharp needles. Write the names of the plants that have needle-like leaves under the heading "Sharp, needle-like leaves" in the chart. Write the other plants under the heading, "Soft, green leaves."

6. Look at the pictures of the animals. Notice that animals have different structures. Write the names of the animals that have wings under the heading "Has wings." Write the names of the other animals under the heading "Doesn't have wings."

7. Complete the chart with the remaining pictures. When you have finished, put the pictures back the way you found them.

8. What traits did you use to classify plants from animals?

____

____

____

9. What other trait could use to classify a rose bush from a maple tree?

____

____

10. What other traits could you use to classify gerbils from fish?

____

____

____

<table>
<tr><th colspan="8">LIVING THINGS</th></tr>
<tr><td colspan="4">Plants:</td><td colspan="4">Animals:</td></tr>
<tr><td colspan="2">Sharp,
needle-like leaves:</td><td colspan="2">Soft, green leaves:</td><td colspan="2">Has wings:</td><td colspan="2">Doesn't have
wings:</td></tr>
<tr><td>Lives in
desert:</td><td>Doesn't
live in
desert:</td><td>Woody
stem:</td><td>No
woody
stem:</td><td>Has
feathers:</td><td>No
feathers:</td><td>Has
backbone:</td><td>No
backbone:</td></tr>
</table>

Woody stem:
- Gives sap for syrup:
- No sap for syrup:

Has backbone:
- Has gills:
- Does not have gills:

Directions (1–21): Each question is followed by four choices. Decide which choice is the *best* answer. Circle the letter of the answer you have chosen.

1 Which place would most likely *not* allow an animal to live and thrive?

A a cave
B the Moon
C a forest
D an open meadow or field

2 If the nutrients found in food are necessary for animals to live and grow, why are waste products produced?

A Animals eat too much.
B Not all of the nutrient material can be used by the animals.
C The animals do not use the nutrients that taste bad.
D The waste is not related to the food animals eat.

3 All starfish can grow back parts of their bodies if they are cut up. This is

A a learned trait
B an inherited trait
C a learned characteristic
D an inherited behavior

4 Which seed would be carried farthest by the wind?

A

B

C

D

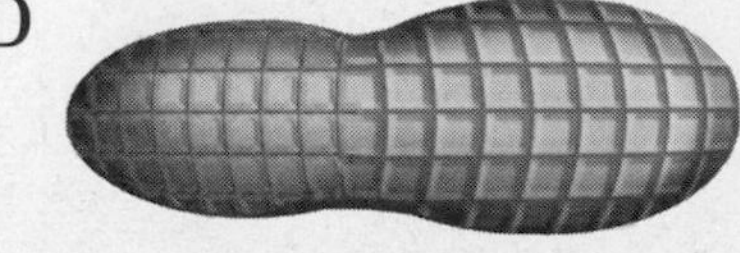

5 Cheryl has brown hair, green eyes, and tanned skin. Which of these traits results from the environment that Cheryl lives in?

A tanned skin
B brown hair
C green eyes
D freckles

6 Which animal does not belong to the same species as the other three?

A German shepherd
B cocker spaniel
C coyote
D poodle

7 Which bird is best able to capture bugs inside tree trunks?

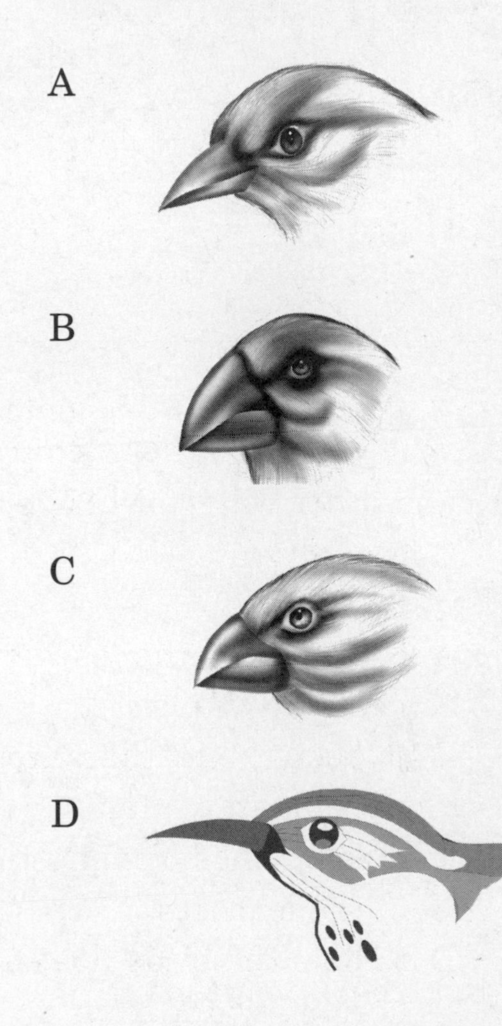

8 Which animals have gills to allow them to live in their environment?

A birds
B fish
C worms
D spiders

9 Which animal has a special and unusual adaptation for eating?

A anteater
B tiger
C turtle
D dog

10 Why do the feathers of ducks and other waterfowl have oils that repel water?

A to help keep the feathers neat and in place
B to keep the feathers from freezing in winter
C to keep bugs from nesting in the feathers
D to help the birds float by keeping water from soaking the feathers

11 Woolly mammoths, ancestors of elephants, had long, thick fur covering their bodies. What was their environment like?

A warm and tropical
B cold and snowy
C warm and dry
D cold and rainy

12 A severe rainstorm has caused half of the flowers to fall off of an apple tree. How will this affect the apples that will grow?

A The apples will be small.
B No apples will grow.
C The apples will be misshapen.
D Fewer apples will grow.

13 Rainwater that falls on the desert soaks into the ground very quickly. Roots must be able to absorb water very fast near the surface. Which picture best shows what cactus roots look like?

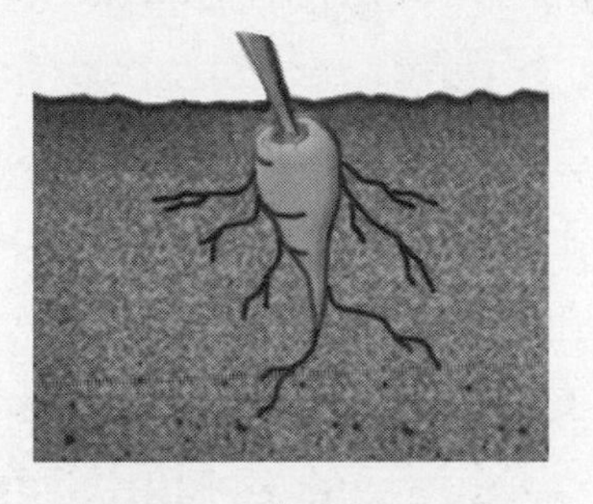

A

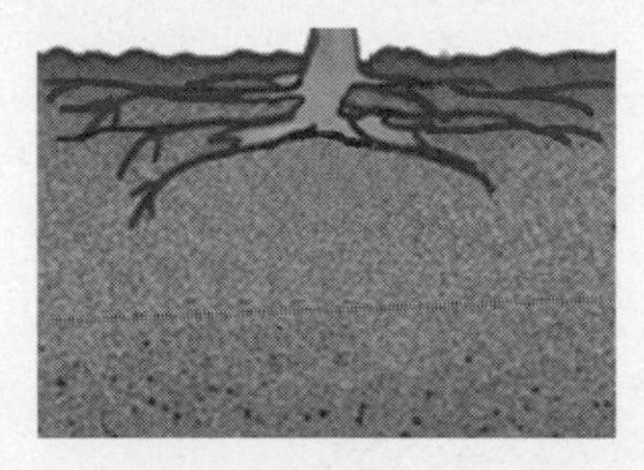

C

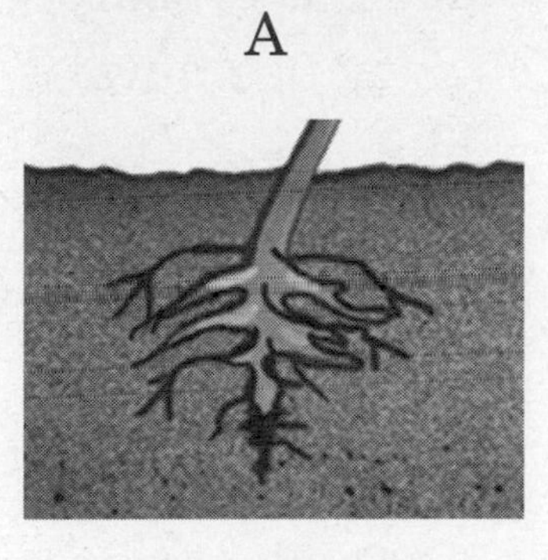

B

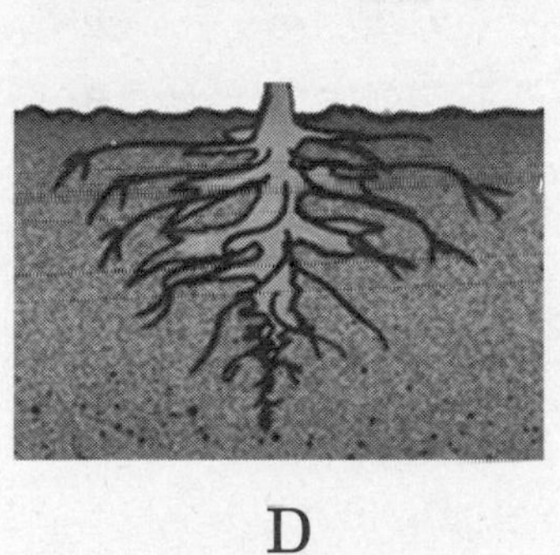

D

14 Which kind of leaf will not absorb sunlight easily?

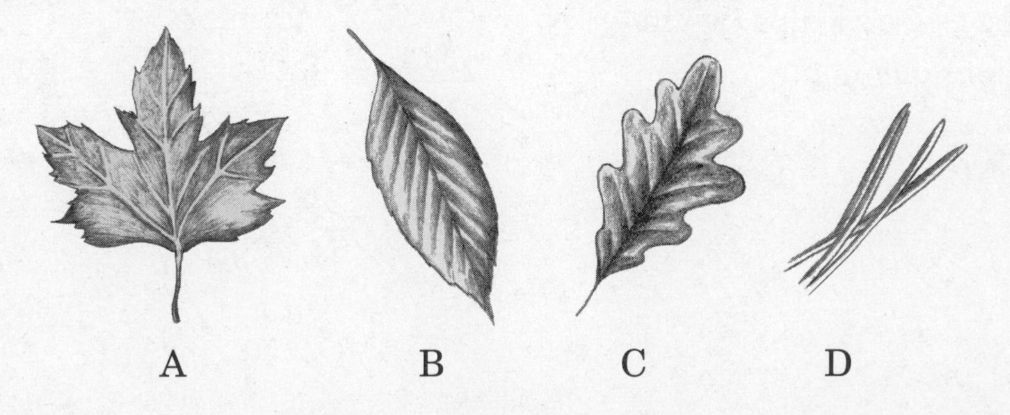

A B C D

15 Sometimes babies cry, just to get picked up. This type of crying is

A a learned behavior
B an inherited trait
C a passed down characteristic from their parents
D an inherited behavior

16 Why are some poisonous plants and animals brightly colored?

A They are in no danger of being eaten by predators.
B The color attracts predators so the plant or animal will kill them.
C The poison helps bright colors develop.
D The color warns predators not to eat them.

17 Why do squirrels collect nuts throughout the fall?

A The nuts prevent them from starving during the winter.
B They can trade food for other items.
C They use them as weapons against predators.
D They use nuts as a nest-building material.

18 What season is it when leaves change color from green to orange?

A winter
B spring
C summer
D fall

19 Your heart is beating very fast. It feels as if it is "pounding." You most likely have been

A surprised
B scared
C angered
D all of the above

20 Plant life requires four items for growth and survival. Which item is *not* available in the deep ocean?

A carbon dioxide
B water
C nutrients
D sunlight

21 Which of the following is a clue for an animal to begin hibernating?

A shorter days and longer, colder nights
B sunny, clear days
C hot days and cool nights
D warm, fair weather

Directions (22–30): For each question, write your answer in the space provided.

22 Cheryl has curly, blonde hair, blue eyes, and freckles. Look at the table below. Which boy is most likely Cheryl's brother? Why?

	ALEX	STUART
Hair	straight, brown hair	curly, red hair
Eyes	brown eyes	green eyes
Skin tone	no freckles	freckles

____

23 How would the migration of birds be affected if there were an unusually warm fall and winter?

____

____

24 Humans perspire when their environment is hot. What other conditions would cause humans to perspire?

____

25 How can a fence made of wood be both a naturally occurring thing and a human-made nonliving thing?

____

26 Milk helps humans grow strong and healthy. Explain whether or not milk would help plants grow healthy too.

____

____

27 Cindy thinks that a hibernating chipmunk is not a living thing because it does not eat. Explain whether she is right or wrong.

____

____

____

28 Why don't some ocean plants have thick trunks?

____

____

29 Some trees have red leaves all year long. You know that plants with green leaves make food for themselves. How do you think the tree with red leaves gets its food?

____

____

____

____

____

30 Some animals have very poor eyesight. Explain which senses you think they use to tell if danger is near.

____

____

Focus on the New York State Learning Standards

Lesson 12 Life Cycles of Plants

4.1a Plants and animals have life cycles. These may include beginning of a life, development into an adult, reproduction as an adult, and eventually death.
4.1b Each kind of plant goes through its own stages of growth and development that may include seed, young plant, and mature plant.
4.1c The length of time from beginning of development to death of the plant is called its life span.
4.1d Life cycles of some plants include changes from seed to mature plant.
4.2a Growth is the process by which plants and animals increase in size.
4.2b Food supplies the energy and materials necessary for growth and repair.
5.2g The health, growth, and development of organisms are affected by environmental conditions such as the availability of food, air, water, space, shelter, heat, and sunlight.

You can describe the stages in the life cycles of plants and learn how environment affects plants.

A **life cycle** is the stages of growth of a plant or animal from the beginning of its life to the end of its life.

A **life span** is the length of time from the beginning of a plant or animal's life until the end of its life.

Guided Instruction

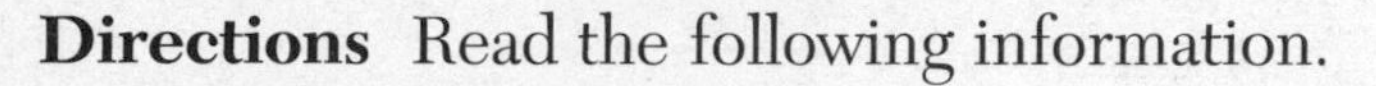

Directions Read the following information.

All plants have **life cycles**. A life cycle includes all the stages of growth from the beginning of a plant's life to the end of its life. This length of time, from the beginning of a plant's development until its death, is called a **life span**.

A plant's life cycle begins as a seed. A seed stores food energy for the tiny embryo plant that is curled up inside the seed. The environment of the seed must be just right to enable the seed to germinate or begin to grow. This will happen when the seed has water, the correct temperature, and air. The seed coat absorbs water and splits to allow a sprout with tiny roots to grow from the embryo.

LIFE CYCLE OF A FLOWERING PLANT

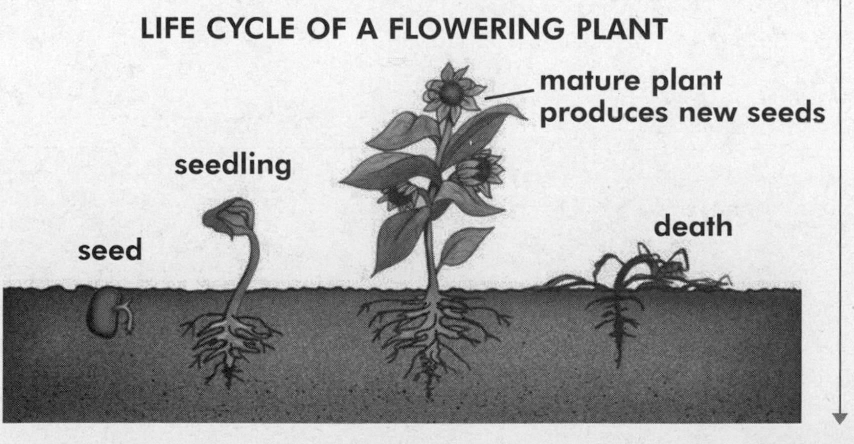

Guided Questions

When does the **life cycle** of a plant begin?

Which do you think would have a longer, natural **life span**, an apple tree or a bean plant?

The tiny plant, called a seedling, develops leaves. The leaves take over the job of making food for the plant. Just as the seed needed the proper environment to sprout, the young plant also needs a proper environment to grow and develop. The young plant needs water, air, and sunlight to make the food it needs. If a plant does not get these things, it will not be a healthy, growing plant.

The healthy plant continues to grow and after it is a mature plant, it will produce flowers and seeds. In fact, the apple you eat holds the seeds that can begin the life cycle of a new apple tree. Apple trees have a longer life span than bean plants that provide the beans you might have on your dinner plate. Apple trees can grow and produce apples each year for twenty or more years. A bean plant has a much shorter life span. It grows and develops from seed to mature plant in only a few months. At the last stage of its life cycle, the plant will wither, turn brown, and die.

Guided Questions

When does the **life cycle** of a plant end?

Directions For each question, write your answer in the space provided.

1. What is the difference between a life cycle and a life span?

____

____

2. What does a seed need in its environment in order to germinate?

____

3. What does a young plant need in its environment in order to grow and develop?

____

____

____

4. The four stages of plant growth are shown above. Write the name of the stage next to the number of each drawing.

1. ____

2. ____

3. ____

4. ____

5. What will a healthy plant produce?

____

____

6. Do all plants have the same life span? Explain your answer.

____

____

____

Apply the New York State Learning Standards to the State Test

Directions: For each question, write your answer in the space provided. Base your answers to questions 7 through 11 on the paragraph below.

Your class is visiting a pumpkin farm in upstate New York. The farmer takes you on a tour of the farm. He explains that last autumn, he plowed the field to get the soil ready for the next spring's planting. This year, New York had a very cold spring, and although the farmer planted the pumpkin seeds on time, they did not sprout right away. He points out the large yellow flowers on the young vines in the field. He says he will hire a beekeeper to bring bees to the field tomorrow to help pollinate the flowers. The farmer tells your class that this fall, he's expecting a late, but plentiful crop.

7 Why do you think the pumpkin seeds were slow to germinate?

____

____

8 At what point in the life cycle are the pumpkin plants the students are viewing? How do you know?

____

____

9 Why is the farmer hiring a beekeeper?

____

____

____

10 At what point in the life cycle will the farmer harvest the pumpkins?

____

11 About how long is the life span of a pumpkin plant?

____

Directions (12–17): Each question is followed by four choices. Decide which choice is the best answer. Circle the answer you have chosen.

12 What is the length of time from the beginning of a plant's development until its death?

A environment
B life span
C seed
D life cycle

13 Which of the following tells the order of the stages of plant development?

A mature plant → seedling → seed
B seedling → seed → mature plant
C seed → seedling → mature plant
D seed → mature plant → seedling

14 Which stage in the life cycle is shown in the drawing?

A mature plant
B seed
C seedling
D dead plant

15 To begin growing, a seed does not need

A water
B correct temperature
C air
D a flower pot

16 Which of the following can begin a new life cycle for an apple tree?

A leaf
B apple
C stem
D root

17 Apple trees can live for many years, but bean plants usually live only a few months. This statement tells us that

A plants have life spans of different lengths
B plants depend on other plants
C plants produce many seeds
D plants bloom in different months

Focus on the New York State Learning Standards

Lesson 13 Life Cycles of Animals

4.1a Plants and animals have life cycles. These may include beginning of a life, development into an adult, reproduction as an adult, and eventually death.
4.1e Each generation of animals goes through changes in form from young to adult. This completed sequence of changes in form is called a life cycle. Some insects change from egg to larva to pupa to adult.
4.1f Each kind of animal goes through its own stages of growth and development during its life span.
4.1g The length of time from an animal's birth to its death is called its life span. Life spans of different animals vary.
4.2a Growth is the process by which plants and animals increase in size.
4.2b Food supplies the energy and materials necessary for growth and repair.
5.2g The health, growth, and development of organisms are affected by environmental conditions such as the availability of food, air, water, space, shelter, heat, and sunlight.

You can describe the stages in the life cycles of animals and learn how environment affects animals.

An animal's **life cycle** is the stages of growth or changes in form from the beginning of its life to the end of its life.

The length of time from an animal's birth to its death is called its **life span**.

Reproduction is the process in which animals produce offspring.

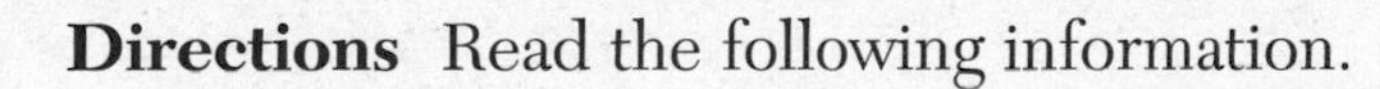

Directions Read the following information.

All animals have **life cycles**. A life cycle includes all the stages of growth from the beginning of an animal's life to the end of its life. The length of time from the beginning of an animal's development until its death is called a **life span**. Life spans of different animals may be just a few days, or as long as 100 years.

Each generation of animals goes through changes in form from young to adult. As the animals change, they grow, or increase in size. Each kind of animal goes through its own stages of growth and development during its life span. Three examples are shown in the chart below.

Guided Questions

What does the **life cycle** of an animal include?

Which has a longer, natural **life span**, a silkworm or turtle?

ANIMAL	LIFE SPAN	CHANGES IN FORM			
insect: silkworm	11 weeks	egg →	larva →	pupa →	adult moth
amphibian: frog	up to 10 years	egg →	tadpole →	tadpole → with tail and gills	adult frog with legs and lungs
reptile: turtle	30 to 100 years	egg →	hatchling →	young turtle →	adult turtle

Food supplies the energy and materials that animals need to grow, develop, and repair their bodies. For example, a well-fed puppy will grow into a healthy adult dog because the food it eats is changed into the energy and materials it needs to develop its body parts.

Reproduction is also part of the life cycle. Reproduction is the process of producing young animals that are the same kind of animal as the mature, parent animals. When the offspring, or young animals, reach adulthood, they, too, will produce their own offspring. This starts the life cycle all over again.

Animals depend on the environment for the food, air, water, space, shelter, and heat that they need. If an animal's habitat, the place where it lives, does not have everything the animal needs, then the animal might not be healthy. It might not grow and develop into a mature, adult animal.

Guided Questions

What does **reproduction** in animals produce?

Directions For each question, write your answer in the space provided.

1. What is the difference between a life cycle and a life span?

2. The four stages of growth of a frog are shown below. Describe what is happening next to the number of each drawing.

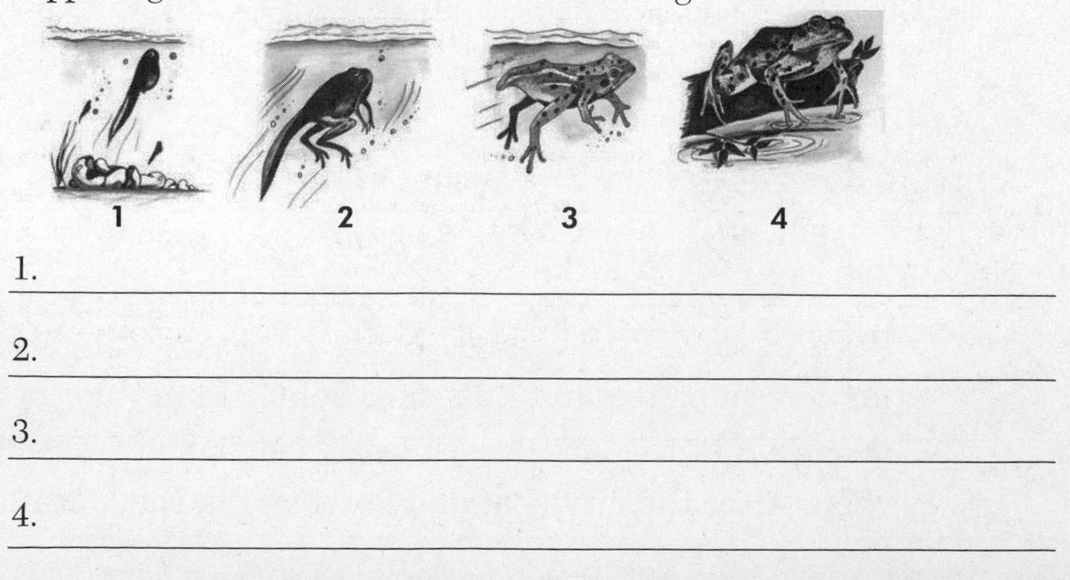

1.

2.

3.

4.

3. What does an animal need in its environment in order to grow and develop?

____

____

4. What is reproduction?

____

____

5. What would the result be if animals did not reproduce?

____

____

6. Do all animals have the same life span? Explain your answer.

____

____

____

Apply the New York State Learning Standards to the State Test

Directions: For each question, write your answer in the space provided. Base your answers to questions 7 through 10 on the paragraph and drawings below.

Your class is raising butterflies so you can observe the life cycle of a butterfly. Make sure that you supply the proper environment for each stage of the butterfly's life from egg to larva to pupa to adult butterfly. Someone in your group should be sure that the larvae, also called caterpillars, have plants to chew. After the larva eat and grow larger, they form a pupa or cocoon. During this stage of development, the adult butterfly forms. When it hatches, the adult butterfly needs sugar water or flower nectar to sip. When you release the adult butterflies, they will lay their eggs on plants where the eggs will lay dormant through the winter. In the spring the eggs will hatch and begin the life cycle over again. After laying eggs, some adult butterflies will fly south toward the warmer weather.

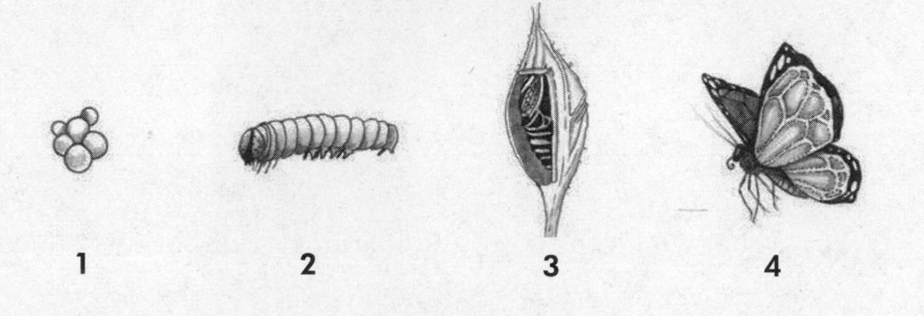

7 What four stages in the butterfly's life cycle did the class observe? Describe what is happening at each stage.

1. ____

____

____

2. ____

____

____

3. ____

____

____

4. ____

____

____

8 At what point in the butterfly's life cycle will the class need to supply plants?

____

9 At what point in the butterfly's life cycle will the class need to supply flower nectar or sugar water?

____

10 What do you think would happen if the released butterflies were unable to fly south? Why?

____

____

____

Directions (11–16): Each question is followed by four choices. Decide which choice is the best answer. Circle the answer you have chosen.

11 What is the length of time called from the beginning of an animal's development until its death?

A environment
B reproduction
C life span
D life cycle

12 Which of the following tells the order of the stages of a turtle's development?

A egg → hatchling → young turtle → adult turtle
B egg → young turtle → hatchling → adult turtle
C hatchling → young turtle → adult turtle → egg
D adult turtle → egg → young turtle → hatchling

13 Life cycles begin again when adult animals

A die
B go south for warmer weather
C reproduce
D grow

14 Which stage in the life cycle of a butterfly is shown in the drawing?

A egg
B caterpillar
C pupa
D adult butterfly

NYS Test Tip

Life Science For some animals, the adult form looks similar to the early form. In others, the two forms are very different.

15 Which of the following statements is false?

A As animals grow and develop, they change.
B All animals go through the same stages of growth and development.
C Life spans of different animals vary.
D The energy that animals need to grow is supplied by food.

16 Which of the following would contribute to the good health of an animal?

A lack of food
B the proper amount of water
C lack of shelter
D air temperatures that are too cold

Focus on the New York State Learning Standards

Lesson 14 Growth of Plants and Animals

4.2a Growth is the process by which plants and animals increase in size.
4.2b Food supplies the energy and materials necessary for growth and repair.

Food supplies the energy and the materials that are necessary for plants and animals to grow.

Growth is the process by which plants and animals get bigger.

Metamorphosis is the process of change in body size and shape of some animals as they develop.

Guided Instruction

Directions Read the following information.

Growth is the process by which plants and animals increase in size. Food supplies the energy and materials that enable plants and animals to grow. All living things need energy to live and grow, but plants and animals get energy and grow in different ways.

Plants grow by turning the Sun's energy into sugar and starches, which they use to make leaves, flowers, and fruits. Plants change some sugars and store them as starches. The sugars and starches that plants use to live and grow might be stored in their roots, stems, leaves, fruits, and seeds. When people and animals eat carrots, which are roots, or tomatoes, which are fruits, or asparagus, which are stems, they are eating the sugars and starches that the plant stored.

Animals need energy to live and grow. They get the energy they need from the food they eat. Some animals, like spiders, crayfish, and termites, shed their hard outer covering, which is called an exoskeleton. The exoskeleton does not grow as the animal grows, so the animal must shed, or molt, its exoskeleton. At each stage of growth, these animals molt to develop a new and larger outside skeleton.

Guided Questions

What does food supply to plants and animals?

How do animals with exoskeletons grow?

Animals with internal skeletons, such as chickens, horses, and turtles, do not molt. The bones inside their bodies grow and they do not change form. They just grow bigger.

Other animals, such as butterflies and moths, go through a process called **metamorphosis**. This means that their bodies change form. First they hatch from the egg as a larva or caterpillar. The larva or caterpillar then eats, grows, and forms a chrysalis or cocoon. Inside the cocoon or chrysalis, the caterpillar is called a pupa. The pupa then changes form and an adult butterfly or moth will emerge.

Guided Questions

What is **metamorphosis?**

METAMORPHOSIS OF A BUTTERFLY

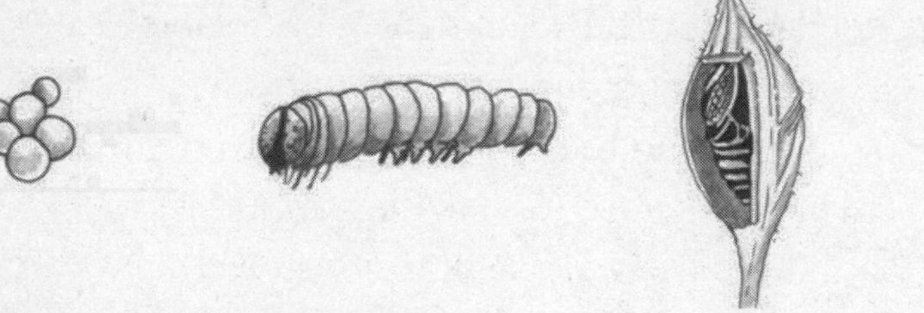

Directions For each question, write your answer in the space provided.

1. What is growth?

____

____

2. What do plants and animals need in order to grow?

____

____

3. You eat plants for food. Name three plant parts that you eat and the plant that the part comes from.

____

____

4. Name three ways animals grow.

____

____

____

5. When and where does a butterfly go through metamorphosis?

____

6. How does having an exoskeleton affect how an animal grows?

____

____

____

____

Apply the New York State Learning Standards to the State Test

Directions: For each question, write your answer in the space provided. Base your answers to questions 7 through 10 on the paragraph and drawing below.

Plants make food to grow in a process called photosynthesis. Here's how they make their own kind of sugar for food:

- The roots take in water from the soil.
- The water goes through the stem to the leaves.
- The leaves take in carbon dioxide from the air.
- The plants get energy from sunlight, which they use to make sugar in the leaves and release oxygen into the air.

7 Why do plants make a kind of sugar?

____

8 What three things do plants need to make sugar?

____

9 What is this process of making sugar called?

____

10 Would a plant be able to grow and live if it did not have one of the three things you named in question 8? Explain your answer.

____

____

____

Directions (11–16): Each question is followed by four choices. Decide which choice is the best answer. Circle the answer you have chosen.

11 The main job of roots in a plant is to

A make food
B reproduce
C take in air
D take in water

12 Where do animals get the energy they need to live and grow?

A from the water they drink
B from the sunlight
C from the food they eat
D from the air they breathe

13 Which is the correct sequence of the growth of a bee?

A egg → cocoon → pupa → adult bee
B egg → larva → cocoon → adult bee
C egg → larva → pupa → adult bee
D pupa → larva → adult bee → egg

14 How do these chicks grow into adult chickens?

A They go through metamorphosis.
B Their bones grow inside their bodies.
C They have an exoskeleton, so they must molt to grow.
D They use photosynthesis to grow.

15 Which of the following is not necessary for the growth of a plant?

A water
B sunlight
C pupa
D carbon dioxide

16 Look at the graph below. It shows how two plants labeled E and F grew over four weeks. Use the data in the graph to choose the statement that is true.

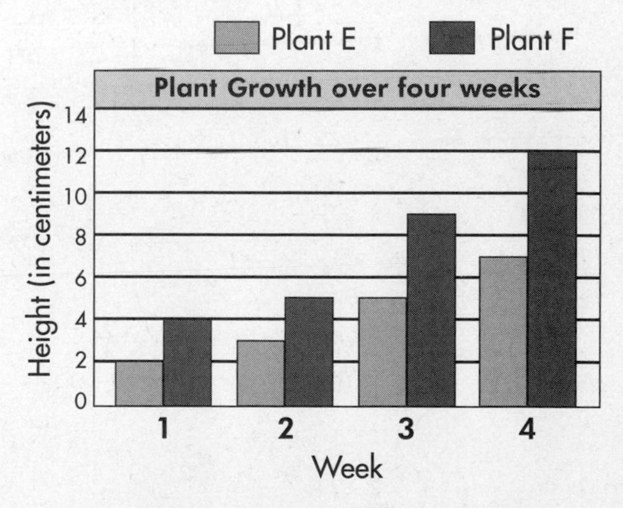

A Both plants grew at the same rate.
B Neither plant grew in four weeks.
C Plant E grew taller than plant F.
D Plant F grew taller than plant E.

Focus on the New York State Learning Standards

Lesson 15 Human Growth and Development

5.3a Humans need a variety of healthy foods, exercise, and rest in order to grow and maintain good health.
5.3b Good health habits include hand washing and personal cleanliness; avoiding harmful substances (including alcohol, tobacco, illicit drugs); eating a balanced diet; engaging in regular exercise.

You can identify good health habits that promote health and growth in humans.

Having a **balanced diet** means eating a variety of healthy foods every day. **Harmful substances** are things such as alcohol, tobacco, and drugs that do damage to a body that will last a lifetime.

Guided Instruction

Directions Read the following information.

These five healthy habits can lead to good health and good growth in people:

- eat healthy foods as part of a **balanced diet**
- exercise regularly
- get enough rest
- have personal cleanliness
- avoid **harmful substances**

A healthy, balanced diet includes eating fruits and vegetables; bread and cereal; meat, fish, and milk products; and a small amount of fats, oils, and sugars every day. Your body needs all these foods because they provide the nutrients that help to keep your body healthy and that help your body fight disease. For example, vitamin C in oranges helps your body fight infection, calcium in broccoli helps you develop strong teeth and bones, carbohydrates in cereal provides energy that your body needs, and protein in beans helps your body grow.

Activities such as swimming or bicycling are good exercises that can help make your body stronger and help your body parts work better. Regular exercise also helps to reduce the risk of many diseases, such as heart disease. Also important for good health is getting enough rest each night.

Good health habits also include good cleanliness habits. Washing your hands with soap and water, and

Guided Questions

What is a **balanced diet?**

What are the benefits of regular exercise?

not sharing towels or drinking glasses, can help you avoid spreading or getting diseases such as colds or flu.

Avoiding harmful substances, like alcohol, tobacco, and drugs will also help you be healthy and stay healthy. Drinking alcohol slows down your nervous system and can cause damage to your liver and other organs. Using tobacco can cause high blood pressure, increase your risk for disease, and do damage to your respiratory system. Abusing drugs can cause damage to your body that can last your entire life.

Guided Questions

What are three examples of **harmful substances?**

Directions For each question, write your answer in the space provided.

1. Make a grocery list of foods that would provide a balanced breakfast.

____

____

2. Name an activity that could provide regular exercise.

____

____

3. How much sleep do you usually get each night? Do you think you get enough rest?

____

4. What is one way people can avoid spreading a cold?

____

5. Why should people avoid harmful substances?

____

____

6. Look at the five habits that can lead to a healthy body. On which one do you think you need to improve? What can you do to improve?

____

____

Apply the New York State Learning Standards to the State Test

Directions: For each question, write your answer in the space provided. Base your answers to questions 7 through 10 on the paragraph and Food Guide Pyramid below.

To eat a healthy, balanced diet every day, start with breads, cereals, rice, pasta, vegetables, and fruits. Then add 2 to 3 servings from the milk group and 2 to 3 servings from the meat group. Remember that the fats, oils, and sweets at the tip of the pyramid should be only a very small part of your diet.

FOOD GUIDE PYRAMID

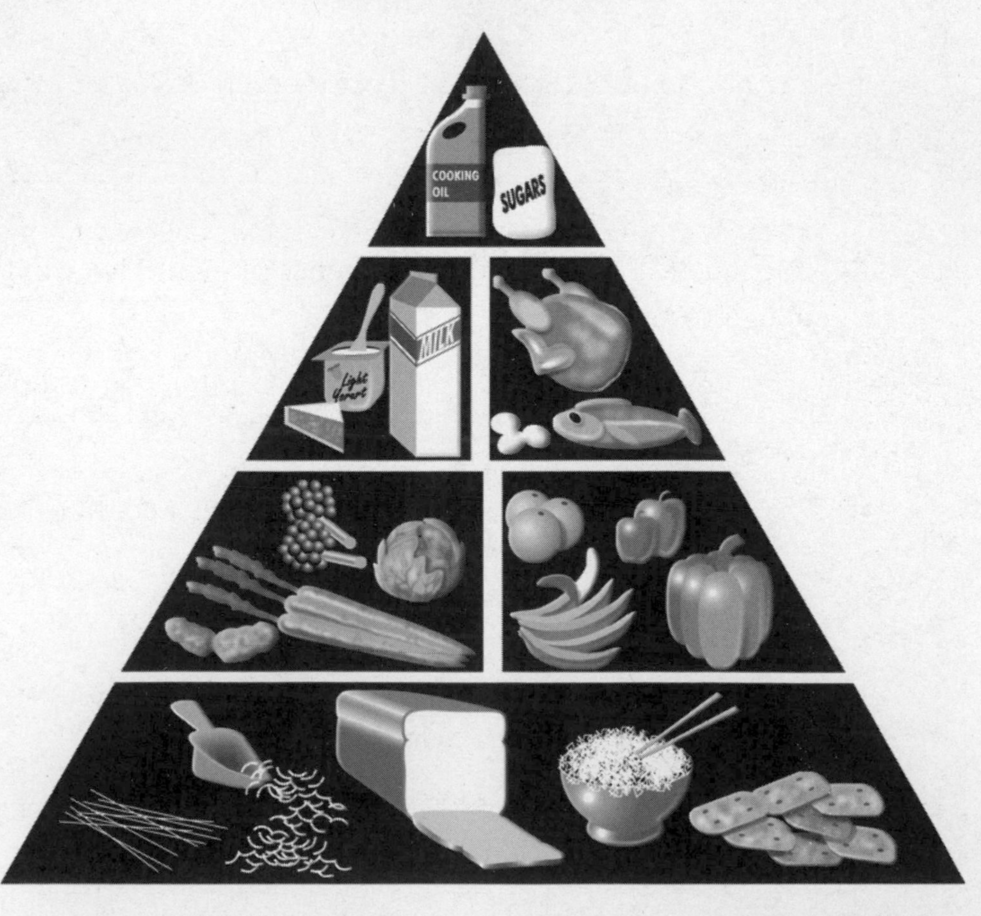

7 In a healthy diet, which of the following should you eat the least: fruits, chicken, candy? Why?

____

____

8 Name a food that would be a healthy snack. In which section of the Food Guide Pyramid does it belong?

____

____

9 Which of the following is the healthiest choice to eat with your lunch: potato chips, carrot sticks, or a candy bar? Why?

____

____

10 Would a dinner of baked chicken, rice, and broccoli with cheese sauce be a balanced meal? Why?

____

____

____

Directions (11–16): Each question is followed by four choices. Decide which choice is the best answer. Circle the answer you have chosen.

11 What does having a balanced diet mean?

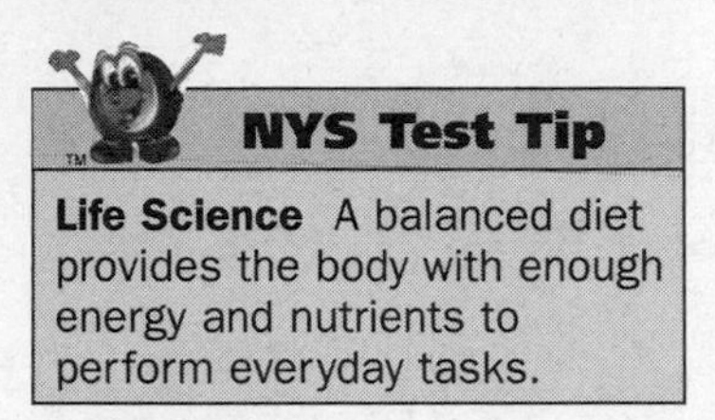

A Each food on your plate weighs the same.
B You have measured or weighed each portion of food.
C You eat a variety of healthy foods every day.
D Everything on your plate is the same color.

12 Which of the following is an example of a harmful substance?

A carrots
B cigarettes
C chocolate cake
D chicken

13 Which of the following would not help you maintain good health?

A eating a balanced diet of healthy food
B riding your bicycle every day
C washing your hands before you eat
D getting two hours of sleep each night

14 Which statement would be the best advice for a friend who eats only at fast food restaurants?

A You should walk to the restaurant.
B You should eat more fresh fruits and vegetables.
C Be sure to ask for French fries with your meal.
D You might want to take a nap after you eat.

Use the Food Guide Pyramid to help you answer questions 15 and 16.

15 Why is the bread, cereal, rice, and pasta group shown on the base of the Food Guide Pyramid and not at the top of the pyramid?

A They are the heaviest foods.
B They are at the bottom of the food chain.
C They make up the largest portion of a balanced diet.
D They are full of fats, oils, and sweets.

16 Which menu best represents a balanced diet?

A glass of milk, an apple, a cheese sandwich
B French fries, potato chips, baked potato
C a slice of white bread, butter, spaghetti noodles
D a banana, an apple, a peach, an orange

Higher-Order Performance Task

Life Cycles

Task:

You will make a model to show the stages in the life cycle of a monarch butterfly. Then you will describe each stage and tell what happens during that stage.

Materials:

- modeling clay or dough
- poster board
- glue
- pictures of each stage in the life cycle of a monarch butterfly
- markers

Directions:

1. Use the materials above to complete this task.
2. Observe the pictures in the life cycle of a butterfly. Use the clay to make a simple model of each picture.

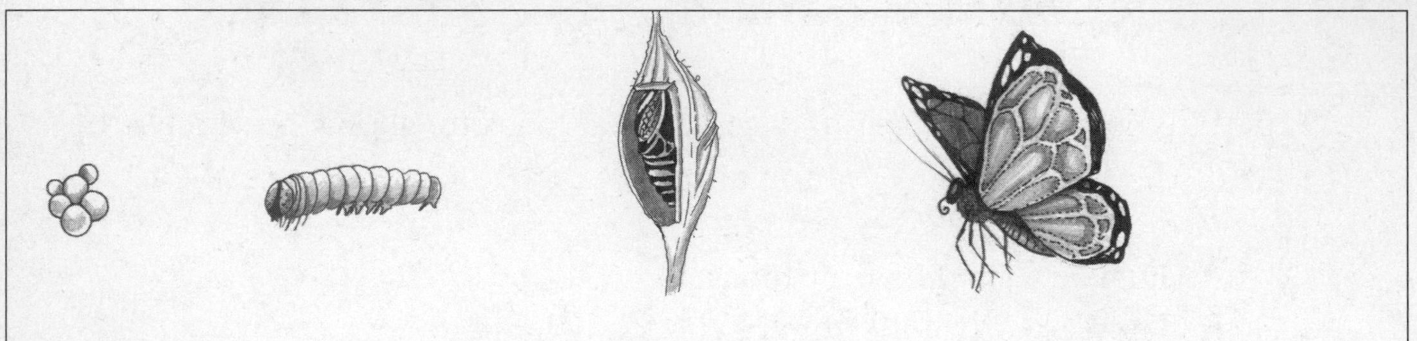

3. Arrange the models you made on the poster board in the correct order. Leave enough space between them to write information. Glue the clay models in place.
4. Use markers to write the name of each stage.
5. Draw arrows to show how the butterfly changes from one stage to the next.
6. Which is the first stage in the butterfly's life cycle? Label it on your poster.

7. What scientific tool could help you find butterfly eggs?

8. During which stage does the butterfly eat and grow? Label it on your poster.

9. What happens during the pupa stage? Label it on your poster.

10. During which stage does the butterfly reproduce? Label it on your poster.

11. Female monarch butterflies lay their eggs only on milkweed plants. When the eggs hatch, milkweed is the only food the caterpillars eat. How will the life cycle be affected if milkweed plants in an area are removed?

12. Use clay to make a model of a milkweed plant. Add it to your poster.

13. Butterflies born in fall travel south where temperatures are warmer. This is known as migration. How do you think migration is important in the life cycle?

14. When you have finished, put the materials away where your teacher instructs you.

Directions (1–20): Each question is followed by four choices. Decide which choice is the best answer. Circle the letter of the answer you have chosen.

1 When you buy a small tomato plant in the spring, which part of its life cycle has already taken place?

A beginning of life
B development into an adult
C reproduction
D death

2 You may recently have experienced a "growth spurt." What does this mean?

A You're growing at a steady rate.
B Your growth rate slowed down for a while.
C Your growth rate sped up for a while.
D You are the average height and weight for your age group.

3 During which season do plants have their greatest increase in size?

A winter
B spring
C summer
D fall

4 Crabs have a hard skeleton on the outside of their bodies. As they grow, they must shed the outer skeleton and grow a new one. What would happen if a crab was not able to shed the old outer skeleton as it grew?

A It probably would die.
B The animal would be the same size forever.
C Its claws would stop growing.
D Its claws would fall off.

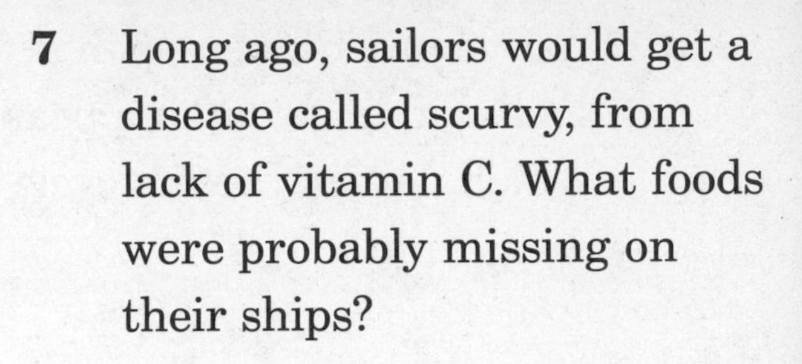

5 When would a crab be most in danger from predators?

A right before it sheds its outer skeleton
B while it is shedding its outer skeleton
C just after it sheds its outer skeleton
D level of danger is always the same

6 You decide to move your healthy, green plant from the kitchen window to a shelf on the kitchen wall. You care for the plant but the leaves change from green to white. What necessary item for successful growth have you changed?

Window

Shelf

A water
B air
C sunlight
D nutrients

7 Long ago, sailors would get a disease called scurvy, from lack of vitamin C. What foods were probably missing on their ships?

A fish
B fruits and vegetables
C cheese
D bread

8 Some caterpillars munch on leaves 24 hours a day. Why do you think they need to eat so much?

A The caterpillar needs to grow so it can be food for birds.
B The caterpillar will soon be a cocoon where it needs energy for metamorphosis.
C The caterpillar will soon lay eggs.
D The caterpillar is near the end of its life cycle.

9 You've just broken open a shell and have eaten a peanut. Which part of the life cycle is the peanut?

A seed to young plant
B young plant to mature plant
C mature plant to death
D death to seed

10 Because alcohol and drugs slow down your thinking and reflexes, which of the following should a person not do if they have been drinking?

A drive a car
B mow the lawn
C ride a bicycle
D all of the above

11 Which of the following will help an animal live a normal life span?

A lots of food available
B possible mates available
C a crowded habitat
D no water available

12 You notice small green sprouts coming out of potatoes in a bag at your home. What stage of its life cycle is a potato?

A seed
B young plant
C mature plant
D death

13 You experiment with two groups of mice. One group is fed breakfast cereal, and the other group is fed raisins. The mice that ate the cereal grew bigger and healthier than the raisin-eating mice. What can you conclude?

A Cereal is not healthy for mice.
B Both the cereal and the raisins are healthy to eat.
C The cereal contains nutrients that the mice need to grow.
D The raisins contain nutrients that the mice need to grow.

14 Two hundred years ago, some children grew up with curved backbones. What might have been the problem?

A They ate too much meat.
B They didn't get the proper nutrition.
C They got too much exercise.
D They didn't go to school.

15 Where is the best place to put plant food?

A on the leaves where it can soak into the pores
B in the soil where it will get soaked up by the roots
C into the stems so it can rise up to the leaves
D all over so the plant is covered in it

16 How does a frog's life change once it reaches the fourth stage in its life cycle?

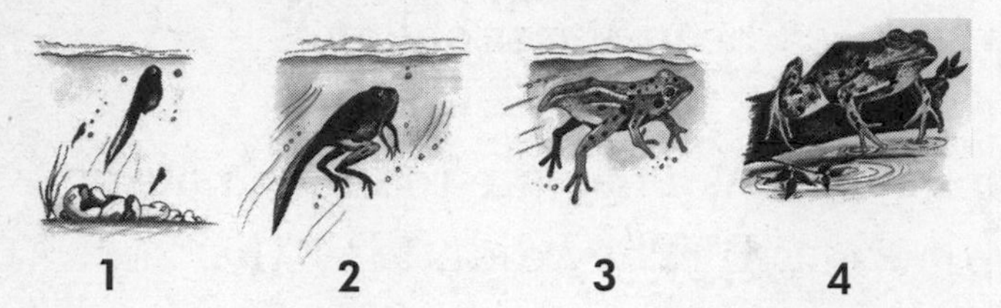

A The frog now lives only on land.
B The frog can now swim long distances.
C The frog can live both in water and on land.
D The frog is no longer a tadpole.

17 In which stage of its life cycle does an average dog spend most of its life?

A beginning of its life
B young adult
C mature adult
D death

18 What part of the human body has the same function as a crab's outer skeleton?

A skin
B hair
C bones
D teeth

19 Susan drinks milk from her friend's carton. Why is this usually thought to be a health risk?

A Susan could get an eye infection.
B Susan could get poison ivy.
C Susan could catch her friend's cold.
D Her friend would get angry if Susan drank all her milk.

20 Alex's father has high blood pressure. What should he do to try to lower it?

A drink more water
B eat a lot of meat
C stop smoking
D sleep less

Directions (21–28): For each question, write your answer in the space provided.

21 Compare the life cycles of a bird and a frog.

Life cycle of a bird:

____

____

____

Life cycle of a frog:

____

____

____

____

22 You have a plant that is getting very large for its flowerpot. How might this affect the plant's growth?

____

____

23 You have a very small fishbowl with six fish in it. Why might this be a problem for the fish?

____

____

24 Why might pizza with green peppers, mushrooms, and tomatoes be considered a healthful food?

____

____

____

25 Annual plants are those we plant each year in the spring. Perennial plants grow again year after year. Which type of plant has a longer life span? Why?

____

____

26 Vitamin D is needed for healthy bones. Humans can use the energy from sunlight to change some nutrients into vitamin D. How is this similar to what plants do?

____

____

____

27 How is metamorphosis in a butterfly different from metamorphosis in a frog?

____

____

28 Fill in five ways to stay healthy in the graphic organizer below.

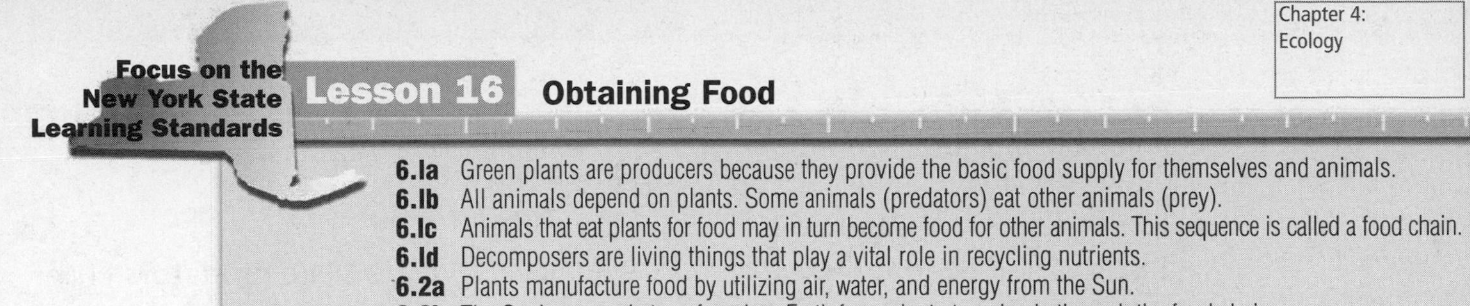

Lesson 16 Obtaining Food

6.Ia Green plants are producers because they provide the basic food supply for themselves and animals.
6.Ib All animals depend on plants. Some animals (predators) eat other animals (prey).
6.Ic Animals that eat plants for food may in turn become food for other animals. This sequence is called a food chain.
6.Id Decomposers are living things that play a vital role in recycling nutrients.
6.2a Plants manufacture food by utilizing air, water, and energy from the Sun.
6.2b The Sun's energy is transferred on Earth from plants to animals through the food chain.

The Sun is the source of energy for life on Earth, and its energy moves through a food chain.

Producers are organisms that use the Sun's energy to make their own food.

Predators are animals that eat other animals for food.

Prey are animals that other animals hunt for food.

A **food chain** is a series of organisms through which energy is passed.

Directions Read the following information.

Most of the energy living things use comes from the Sun. Plants make their own food using air, water, and light energy from the Sun. Roots bring in minerals and water from the soil to help the plant make food. Then the plants store the food and energy for use later when they need it for reproduction and other life processes.

Although other living things cannot make food from the Sun's energy, they use the food made by plants. Green plants are called **producers**. Plants are called producers because they produce the basic food supply for themselves and for all animals.

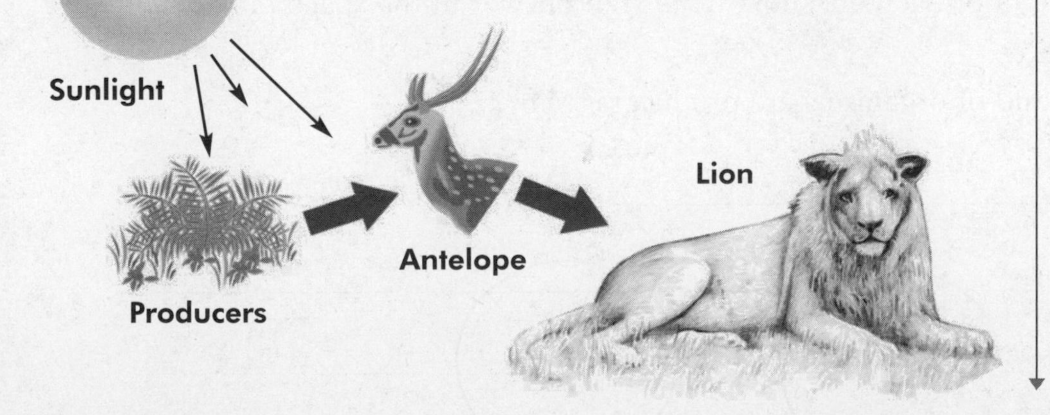

Guided Questions

Why are plants called **producers?**

All animals depend on plants, whether they eat them or not. Some animals eat plants and use the food the plants make from the Sun's energy. Other animals, called **predators,** eat animals who have eaten the plants. For example, after an antelope eats grass, it might become **prey** for a lion. So animals that eat plants for food may in turn be used as food for other animals.

As plants make energy and animals eat plants or other animals, the Sun's energy is passed from plants to animals. Every organism is part of a **food chain** through which energy is passed. A food chain begins with a producer, such as a tomato plant that makes food from the Sun's energy. Next, a tomato worm might eat the tomato leaf and get energy from the food stored in the leaf. A toad might eat the food stored in the tomato worm's body. Then the toad might become a meal for a hungry snake.

Decomposers are the last step in a food chain. These organisms break down the dead bodies of other organisms. Many of these organisms are too small to see without a microscope. Decomposers play important roles. Without them, dead organisms would pile up. Their stored energy and nutrients would be wasted. When decomposers break down the bodies of dead organisms, they return energy and nutrients to the soil for plants to use.

Guided Questions

How are **predators** and **prey** related?

What is a **food chain?**

How do decomposers recycle energy?

Directions For each question, write your answer in the space provided.

1. What kind of organisms are producers? Why?

____

____

____

2. Give three examples of predators.

3. What type of animal might be prey for a fox?

4. How do predators get energy from the Sun?

5. Why doesn't the soil run out of nutrients?

6. When people say "eat low on the food chain," what type of food are they recommending?

Apply the New York State Learning Standards to the State Test

Directions: For each question, write your answer in the space provided. Base your answers to questions 7 through 12 on the drawing below.

7 What organisms in the picture are producers?

____

8 What organisms in the picture eat plants?

____

9 What organism in the picture is a predator?

____

10 What organism in the picture could be prey for the predator?

____

11 Make a food chain diagram using three organisms in the picture.

12 What would probably happen if something killed all the producers in the area?

____

Directions (13–18): Each question is followed by four choices. Decide which choice is the best answer. Circle the answer you have chosen.

13 The Sun is a source of energy for

A predators
B prey
C both predators and prey
D neither predators nor prey

14 What do producers need to make food?

A air, water, and energy from the Sun
B fertilizer, water, and energy from the Sun
C water and energy from the Sun
D air and energy from the Sun

15 How do animals get energy from the Sun?

A from the air they breathe
B from the water they drink
C through their skin
D from eating plants

16 If all of these organisms were on one food chain, which would be last?

A predator
B decomposer
C prey
D plant

17 Which organism is a producer?

A clam
B maple tree
C honey bee
D cow

18 Suppose a mountain lion's main food source is mountain goats. Mountain goats eat plants. Most of the plants in the area die from lack of rain. The number of mountain lions in the area would probably

A increase
B decrease
C remain the same
D increase and then decrease

Focus on the New York State Learning Standards

Lesson 17 Environmental Change

6.1e An organism's pattern of behavior is related to the nature of that organism's environment, including the kinds and numbers of other organisms present, the availability of food and other resources, and the physical characteristics of the environment.
6.1f When the environment changes, some plants and animals survive and reproduce, and others die or move to new locations.
6.2c Heat energy from the Sun powers the water cycle.

Plants and animals, including humans, depend upon each other and the nonliving environment.

A **resource** is something living things need, such as food, air, or water.

The **water cycle** is the constant movement of water from the ground, to the sky, and back again.

Precipitation is water falling from the sky as rain, snow, sleet, or hail.

Directions Read the following information.

An organism's structure and behavior are adapted, or adjusted, to its environment. For example, characteristics of the land determine what species can live on it, and how many of each species. Land with little heat and water will not support many plants.

Living things also adapt to and depend on the kinds and numbers of other organisms in an environment. If the land has very few growing plants, then very few animals can live there. Whether a species can survive, and how many members can survive, depends on the amount of food and **resources** in the environment. If the environment changes, some plants and animals survive and reproduce. Others die or move to new locations.

Guided Questions

A volcano changed the environment in the picture. What will happen to the plants and animals?

What are **resources?**

Suppose a warm area has plenty of water, grass, and short plants, but not tall trees. Many grass-eating animals, such as deer, will live there. Predators who hunt the grazing animals will live there too. Over time, forests sometimes grow in a grassy area. Grass cannot grow in the shade of the trees, so much of the grass dies. Many grass-eating animals such as deer will die or move to other areas. Shade-loving plants such as ferns may grow on the ground, and tree animals like squirrels may move into the area.

The resources in an environment depend heavily on the Sun. Besides supplying heat and the energy plants need, the Sun's energy powers the **water cycle**. Without heat from the Sun, all water on the ground would soak into Earth or stay in lakes and oceans. It would not fall on land as **precipitation** for plants and other organisms.

Before rain or snow can fall, water needs to rise into the air. The heat of the Sun causes water to evaporate. It then rises high above Earth, where it forms clouds. Rain, snow, hail, or sleet fall to Earth from the clouds.

Guided Questions

What is the **water cycle?**

What is **precipitation?**

Directions For each question, write your answer in the space provided.

1. How could the amount of precipitation affect the number of organisms living in an environment?

2. Give two examples of resources living things need.

3. How does the Sun power the water cycle?

____

____

4. What happens to plants and animals if the environment changes?

____

____

5. What might happen to a grassy plain if a large number of trees grew there?

____

____

Apply the New York State Learning Standards to the State Test

Directions: For each question, write your answer in the space provided. Base your answers to questions 6 through 10 on page 121 on the flow chart below.

Stage 1 A fire burns a forest to the ground.
Stage 2 Grasses and weeds grow in burned area.
Stage 3 Bushes and short trees begin to grow. Because they shade the grasses, some of the grasses die.
Stage 4 Trees begin to grow in the area. Because the trees shade parts of the area, some shrubs die.
Stage 5 A dense forest of crowded trees forms.

6 What would you predict about animals during Stage 2?

7 What would you predict about ground plants during Stage 5?

8 What would you expect to happen to young trees after the mature trees have grown tall and crowded?

9 What would you expect to happen to animal species as the plant species changed from one stage to another?

10 Explain why the forests that are replanted after being cut down are not always the same as the ones that were cut.

Directions (11–15): Each question is followed by four choices. Decide which choice is the best answer. Circle the answer you have chosen.

11 What is the constant movement of water from the ground, to the sky, and back again called?

A rain
B snow
C water cycle
D rain clouds

12 What could cause a drop in the number of wild horses in an environment?

A a decrease in the number of predators
B a decrease in the number of ground plants
C a decrease in the number of grazing deer
D a decrease in the number of people

13 Water falling from the sky as rain or sleet is called

A water cycle
B precipitation
C snow
D hail

14 If many trees are planted on a grassy plain, over time the amount of grass will

A increase
B decrease
C remain the same
D change by turning greener

15 The amount of rain in an area will not affect

A the amount of space for animals to live
B the kinds of plants that live there
C the kinds of animals that live there
D the number of plants and animals that live there

Focus on the New York State Learning Standards

Lesson 18 Humans and Their Environments

7.Ia Humans depend on their natural and constructed environments.
7.Ib Over time humans have changed their environment by cultivating crops and raising animals, creating shelter, using energy, manufacturing goods, developing means of transportation, changing populations, and carrying out other activities.
7.Ic Humans, as individuals or communities, change environments in ways that can be either helpful or harmful for themselves and other organisms.

Human decisions and activities have produced a major impact on the physical and living environments.

Pollution is harmful substances that damage the air, water, land, or food supply.

To **manufacture** is to make something.

Guided Instruction

Directions Read the following information.

Like other organisms, humans depend on their natural environments. Earth provides the energy, nutrients, air, food, water, and heat that humans need. Because they can build part of their environment and move resources from place to place, humans can live in almost any natural environment on Earth. Humans live in frozen areas near the South Pole, in hot dry deserts, and on ships at sea.

Humans build shelters, grow food, **manufacture** goods, and create heat from stored or transported energy sources. The shelters become part of their environment, protecting them from harsh weather and predators. The food they produce and preserve allows them to eat in places with no foods nearby.

Guided Questions

What do humans do when they **manufacture?**

Why are humans able to live in this harsh environment?

The goods they make help them survive conditions such as cold, storms, or dryness. By developing means of transportation, they can bring food, water, energy, and goods to where they need them.

Over time, humans have changed their natural environments and created new ones. Forests and grasslands that once stretched for miles have been replaced by miles of farms and ranches. Near Los Angeles, California, you can ride for almost one hundred miles without seeing anything but towns and cities. Building shelter, using energy, manufacturing goods, using transportation, and other human activities have created **pollution** in the air, land, and water.

Humans sometimes change environments in helpful ways. For example, humans have brought water to dry lands so that plants can grow. They sometimes feed wild animals when food is scarce. Some humans plant trees to replace those that have died from natural causes.

Even so, many changes are harmful to humans and other species. Many species of animals have disappeared as humans hunted them, or replaced their homes with farms or cities. Pollution from human activity makes the land, water, and air less healthy for all organisms. Every day, humans make decisions that affect the environment for better or worse.

Guided Questions

What is **pollution?**

Directions For each question, write your answer in the space provided.

1. Explain how humans created the environment where you live.

____

____

____

____

2. Why can humans live in so many different natural environments, including very hot or very cold ones?

3. Give one example of how humans have changed environments in harmful ways.

4. Give one example of how humans have changed environments in ways that are helpful to humans but harmful to animal and plant species originally living there.

5. Give one example of a decision an individual might make that is helpful to the environment, and one that could be harmful.

Apply the New York State Learning Standards to the State Test

Directions: For each question, write your answer in the space provided. Base your answers to questions 6 through 9 on the drawing below.

The drawing below shows an oil rig drilling for crude oil. It also shows a refinery that processes the crude oil into home-heating oil, gasoline, and many other products such as kerosene and jet fuel. And it shows the trucks that deliver the products to homes and businesses. Humans depend heavily on these products for heating, cooling, and transportation.

6 In what ways could the smoke stacks and the trucks cause pollution in the air?

____

____

____

____

7 In what ways could the refinery or the drilling rig cause pollution in the land?

____

____

____

8 How could the smoke stacks and the drilling rig cause pollution in the water?

____

____

____

____

____

9 What could people do to help lessen the pollution caused by the production and transportation of crude oil products?

____

____

____

____

Directions (10–15): Each question is followed by four choices. Decide which choice is the best answer. Circle the letter of the answer you have chosen.

10 A polar bear could not live in the jungle. An alligator could not live near the North Pole. Humans can live in both places because

A their bodies are adapted to different conditions
B they are naturally stronger than most animals
C they build environments for themselves
D they use more resources more than most species

Strategy If a choice finishes a sentence, try reading the entire sentence to decide if it makes sense.

11 Wherever they are, most humans have the food they need because

A they can grow food anywhere
B they are good at using the natural food sources in the environment
C they can make artificial food with no natural resources
D they can preserve and transport food

12 Which living things are harmed by human-made pollution?

A plants
B animals
C plants and animals
D plants, animals, and humans

13 Which invention would *most* likely be helpful in a cold environment?

A 

B 

C 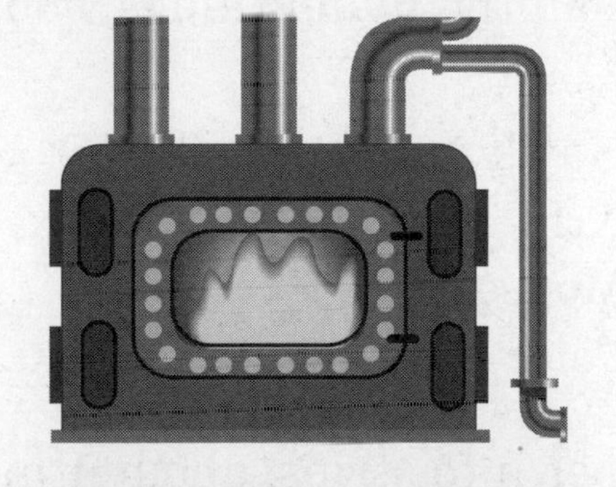

D 

14 Which decision could humans make that would help the environment the most?

A Use electricity instead of gas.
B Eat more manufactured foods and less farmed food.
C Use less energy and fewer goods.
D Care for beautiful gardens.

15 A town near the wilderness is growing. Every month, wilderness is turned into rows of homes. As the town grows, bears are coming into town more often, stealing food from garbage cans. Why are bears *most likely* doing this?

A As bears see more people, they like them better.
B They are losing their natural food sources from the land.
C Bears have discovered garbage tastes better than their normal food.
D It is easier to steal food from garbage cans than to hunt and look for their natural food.

Higher-Order Performance Task

Investigating a Food Chain

Task:

You are going to organize living things into food chains. Then you are going to make branches off your chain to show how food chains can overlap.

Materials:

- construction paper (red, green, and blue)
- scissors
- stapler or tape
- markers

Directions:

1. Look at the list of living things below. Find four living things that might form a food chain. Make sure all the living things would live in the same ecosystem.

Arctic cod	Diatom	Hawk	Rabbit
Arctic wolf	Fungi	Krill	Red fox
Bacteria	Grass	Mouse	Seal
Deer	Grasshopper	Polar bear	Tree

2. Cut three sheets of construction paper into strips, about 1 inch wide. Use a green sheet, a red sheet, and a blue sheet of paper.
3. Write the name of a producer on a green strip.
4. Bend the producer strip into a circle and staple the ends together. This is the first link in the food chain.
5. Write the name of consumers on blue strips.
6. Choose a consumer that feeds on the producer. Bend its strip through the producer link. Staple the ends to add the consumer as a link in the chain.
7. Choose a consumer that feeds on the first consumer. Add it as a link in the chain.
8. Write the name of a decomposer on a red strip. Add it as a link in the chain.

9. Draw the food chain you made in the space below.

10. What producer did you choose? What is its role of in the food chain?

____

____

11. What is the first consumer you chose? What is the second consumer?

____

____

12. What decomposer did you choose? What is its role of in the food chain?

____

____

13. What does the food chain show?

____

14. Look for another consumer that feeds on your producer. Build another food chain from the same producer.

15. Add links to show if a consumer from one food chain feeds on a consumer from the other food chain. When you are finished, you will have a food web. A food web shows food chains that overlap.

16. When you have finished, present your food web to the class and put things back the way you found them.

Building Stamina®

Directions (1–21): Each question is followed by four choices. Decide which choice is the best answer. Circle the letter of the answer you have chosen.

1 If a plant makes more food than it uses, what do you think the plant will do with the extra food energy?

A Store it for future use.
B Release it into the air.
C Grow larger leaves.
D Stop photosynthesis.

2 Why do ducks have webbed feet?

A It makes swimming easier.
B It makes walking easier.
C They help ducks build nests on muddy shores.
D They keep the duck from sinking in water.

3 Panda bears eat only bamboo. The amount of wild bamboo is decreasing because the land is being used for houses and farms. What is happening to the pandas?

A They are learning to eat something else.
B They are in danger of dying out.
C They are moving to different areas.
D They are attacking the construction workers.

4 How does fertilizer from farm crops add to water pollution?

A The fertilizer soaks into the ground.
B The fertilizer is washed by rain into the rivers.
C Fertilizer goes into drinking water when we wash fruits and vegetables at home.
D Wind blows the fertilizer into lakes and oceans.

5 When is an animal both predator and prey?

A when it produces its own food
B when the animal eats every animal it finds
C when it eats some animals but can also be eaten by other animals
D when an animal eats only plants

6 During the summer, why is it often hotter in New York City than it is in the surrounding towns?

A The big buildings block the cooling breezes.
B The streets and buildings absorb heat and release it back into the air.
C There are fewer cars and buses.
D There are more people adding their body heat to the already warm air.

7 Because the amount of energy decreases with each link on a food chain, a large predator needs to

A eat a lot to get enough energy for growth
B eat small amounts to get enough energy for growth
C become a producer
D move fast to save the energy it has

8 Which of the following could be called decomposers?

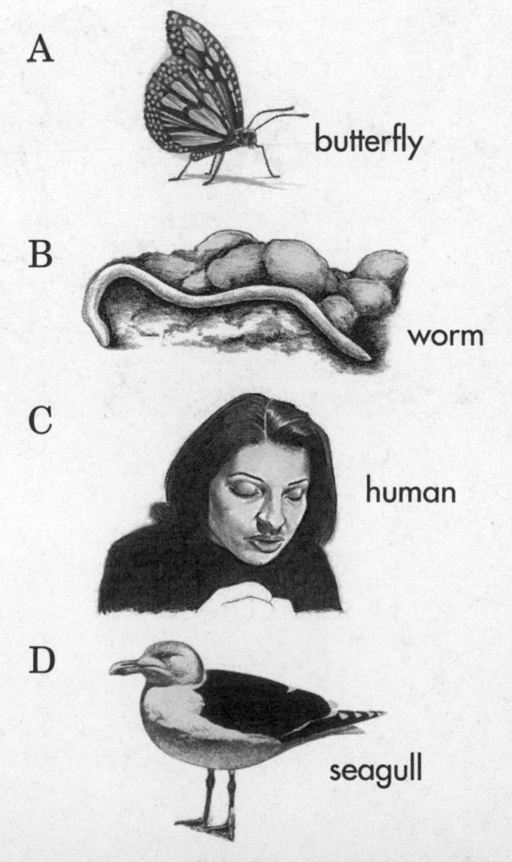

9 Oil spills can be cleaned up with the help of very small organisms that eat the oil. These organisms are

A producers
B predators
C decomposers
D prey

10 Where does the water come from that plants use in making energy from the Sun?

A rain
B groundwater
C water vapor
D all of the above

11 Suppose that certain rocks in the desert contain fossils of pinecones and pine needles. You can infer that

A there were a lot of trees long ago
B this desert was once even drier and hotter than it is now
C this environment was once moist and cooler
D fossil are common here

12 On a warm sunny day, the water level in a pond has gone down. What part of the water cycle has taken place?

A precipitation
B condensation
C liquification
D evaporation

13 A fire has killed most of the trees in a forest. Which animals would most likely return to this area first?

A tree-living animals such as squirrels and birds
B those tree-living animals than can live on the ground
C animals that eat grass, such as deer
D all of the animals

14 Why does the leaf frog actually look like a leaf?

A It can soak up the Sun's energy and make its own food.
B It fits more comfortably in a tree.
C It is camouflaged and safer from predators.
D It can sneak up on prey more easily.

15 Large numbers of prairie dogs live together in "towns." What might it mean when one starts barking and then the rest begin to bark?

A time to eat
B approaching danger
C change in weather
D bedtime

16 Which is not a producer?

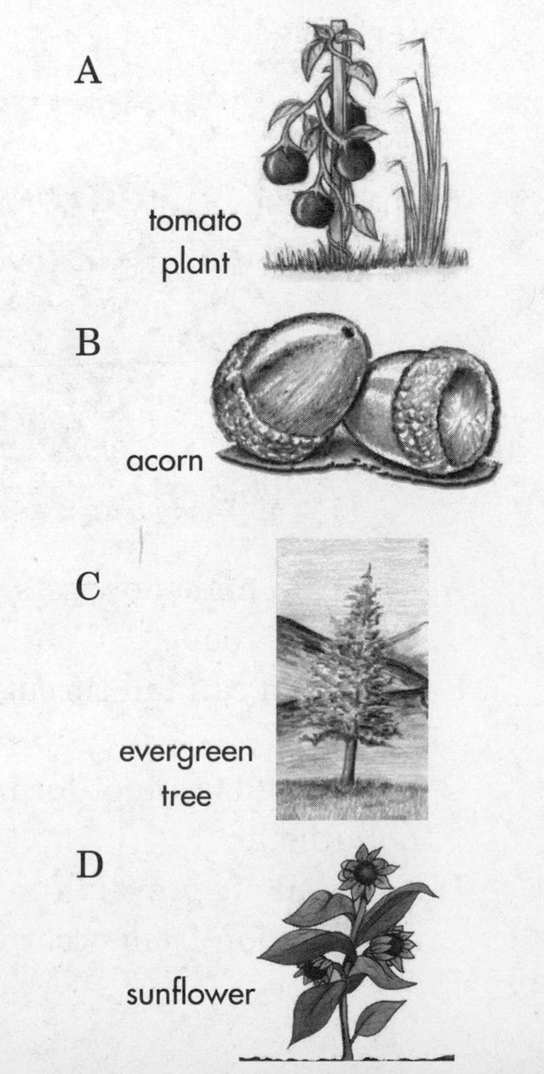

17 If you moved to the North Pole, which of the following would you *not* have to bring?

A food
B water
C air
D heat

18 Look at the diagram. How might polluted air affect the water cycle?

A The rain and snow gets polluted, too.
B Less rain can fall through the air.
C Rain clouds cannot form in polluted air.
D Polluted air prevents evaporation from occurring.

19 Which is *not* a method that the early pioneers used to keep their food from spoiling?

A drying
B freezing
C salting
D freeze-drying

20 A strange plant disease has affected an area and many plants have died. How will this affect this area's food chain?

A The rest of the plants will die.
B The soil will become poisoned by the dead plants.
C The food chain will become smaller.
D The animals that eat those plants may die.

21 How does the building of a water reservoir help a community?

A It provides a never-ending supply of fresh water.
B It may prevent flooding by holding extra water.
C It serves as an area for boating.
D all of the above

Directions (22–27): For each question, write your answer in the space provided.

Base your answers to question 22 on the drawing below.

22 This drawing shows two food chains. Add arrows to show how you could change this into a food web.

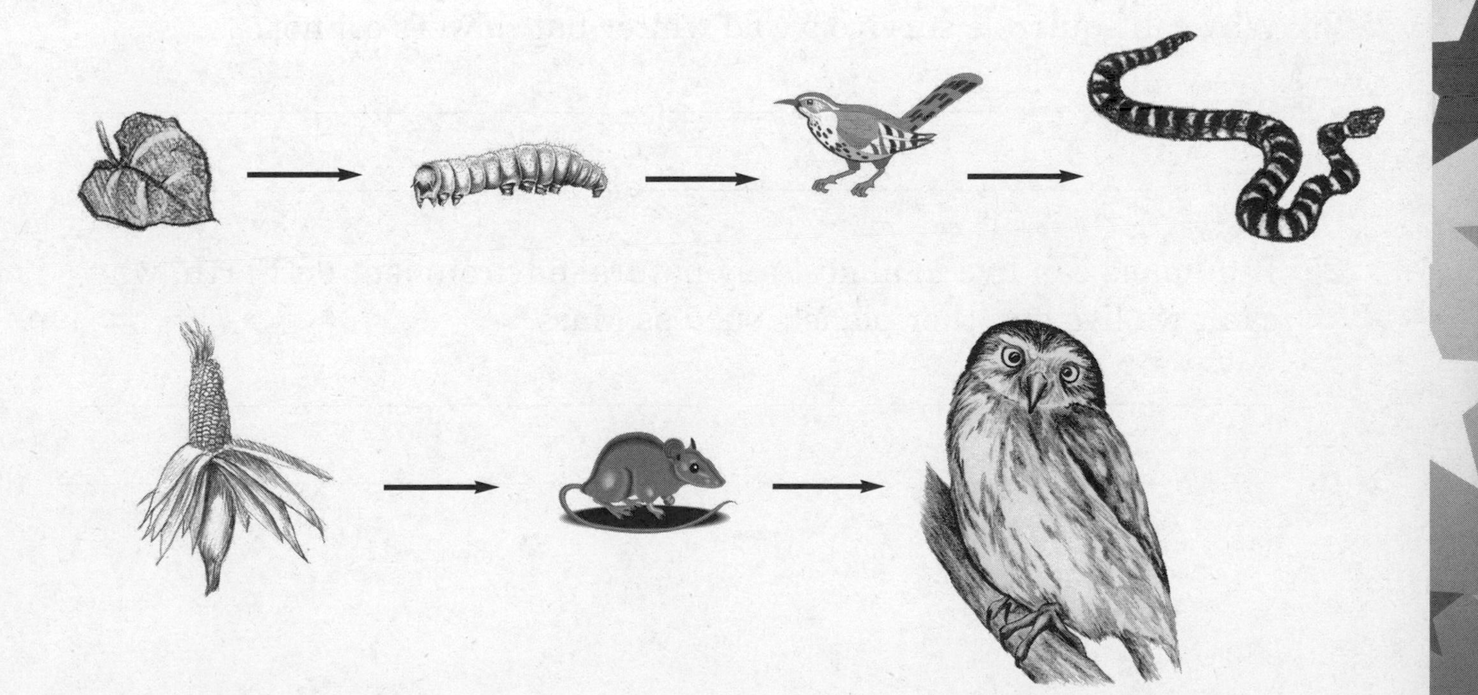

23 Why does a cactus have a fleshy, water-filled body?

____

____

24 Why is soil erosion worse on farmlands than in forests?

____

____

____

25 Why would a metal shack be an unsafe home in the desert?

____

____

26 Why can squirrels survive a cold winter but flowers cannot?

____

____

27 If humans can live in almost any natural environment on Earth, why can't we live on other planets such as Mars?

____

____

Focus on the New York State Learning Standards

Lesson 19 Weather

2.1a Weather is the condition of the outside air at a particular moment.
2.1b Weather can be described and measured by: temperature, wind speed and direction, form and amount of precipitation, general sky conditions (cloudy, sunny, partly cloudy).

Weather can be warm or cold, wet or dry, cloudy or clear, and windy or calm.

Weather is the condition of the outside air at a particular moment.

Temperature is a measure of how warm or cold something is.

Wind is moving air.

Guided Instruction

Directions Read the following information.

What kind of **weather** is outside today? To answer that question, you need to think about the condition of the air at this moment. Is it warm or cold? Is the air dry or wet? Is rain, sleet, hail, or snow falling? Is the air calm and still, or moving fast to make wind? Is the upper air clear, or full of clouds?

If you know the answer to one of these questions, you may know the answer to some of the others. For example, if there is precipitation, you know the sky is cloudy. Precipitation usually falls from dark, thick clouds. If snow is falling, you know the air is cold.

When you watch a weather report on television, what do the reporters tell you? They probably report several measures of the weather. One measure is **temperature**. Temperature depends partly on how much of the Sun's energy is reaching the ground. In the United States, most summer days are warmer than most winter days.

Guided Questions

What are some ways to describe the **weather?**

What is **temperature?**

Moving air is called **wind**. If wind is blowing from a cold place, it will affect the temperature of the air. Sometimes wind blows different weather into an area. Have you heard the expression "A storm is approaching from the west?" Weather experts determine the direction of the wind and measure its speed.

A weather report will include measures of how wet or dry the air is. It tells whether rain, snow, sleet, or hail is falling, and if so, how much in a given amount of time. The reporters describe the sky. If it is cloudy, they can tell you the type of cloud. The type of cloud can tell weather reporters whether or not precipitation is on the way.

Guided Questions

If **wind** is blowing from a cold place, how will it change the weather?

Directions For each question, write your answer in the space provided.

1. What is weather?

____

2. Give six examples of words that could be used to describe the weather.

____

____

3. What is temperature?

____

4. Name four types of precipitation.

____

5. If you look out the window and see rain falling, what does that tell you about the sky?

____

6. Suppose you are going to an outdoor soccer game in another town. You call a friend to ask about the weather there. List four weather questions you should ask.

____

____

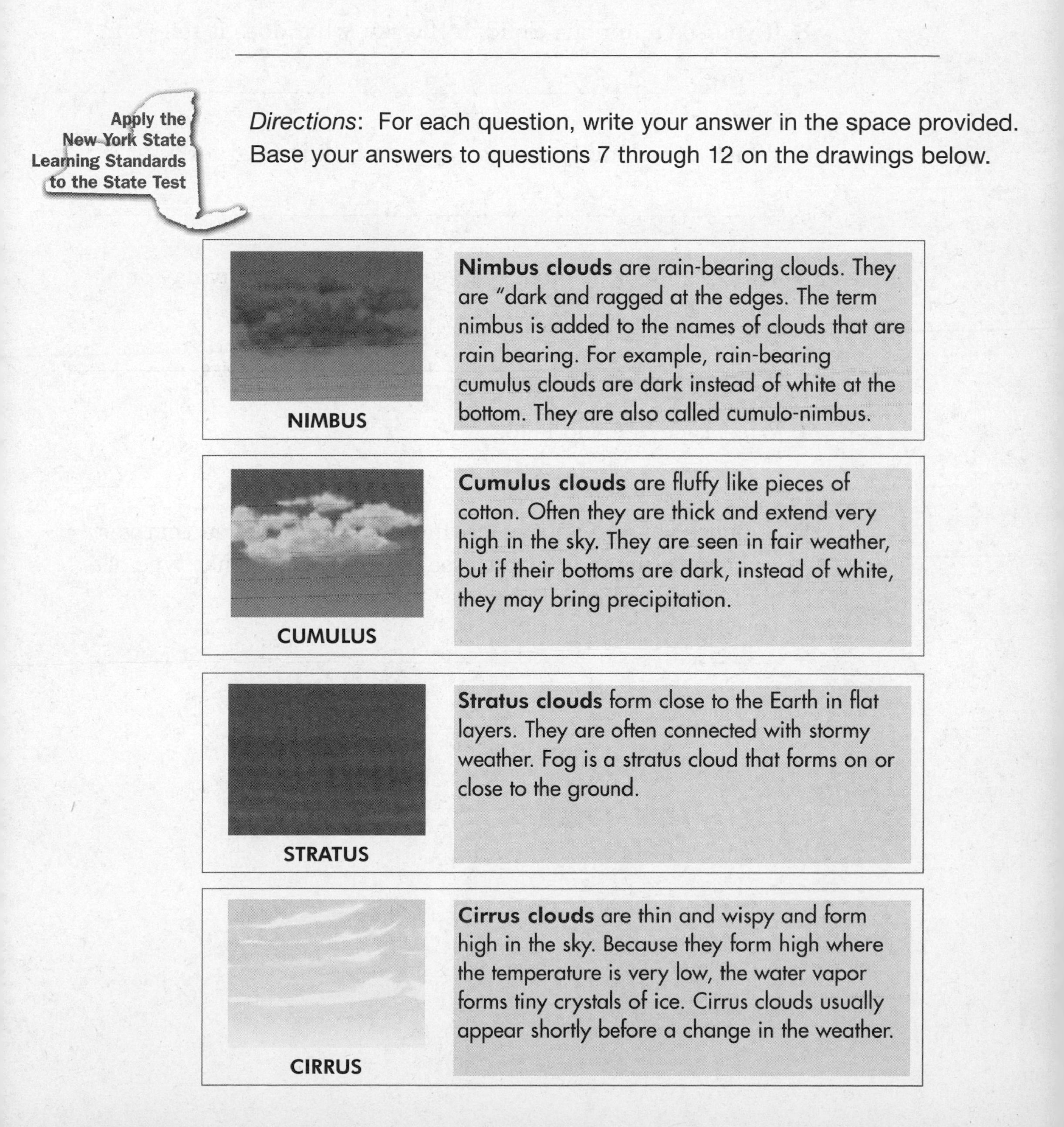

Apply the New York State Learning Standards to the State Test

Directions: For each question, write your answer in the space provided. Base your answers to questions 7 through 12 on the drawings below.

NIMBUS

Nimbus clouds are rain-bearing clouds. They are "dark and ragged at the edges. The term nimbus is added to the names of clouds that are rain bearing. For example, rain-bearing cumulus clouds are dark instead of white at the bottom. They are also called cumulo-nimbus.

CUMULUS

Cumulus clouds are fluffy like pieces of cotton. Often they are thick and extend very high in the sky. They are seen in fair weather, but if their bottoms are dark, instead of white, they may bring precipitation.

STRATUS

Stratus clouds form close to the Earth in flat layers. They are often connected with stormy weather. Fog is a stratus cloud that forms on or close to the ground.

CIRRUS

Cirrus clouds are thin and wispy and form high in the sky. Because they form high where the temperature is very low, the water vapor forms tiny crystals of ice. Cirrus clouds usually appear shortly before a change in the weather.

7 If you saw cirrus clouds after several days of warm weather, what would you predict about the weather?

____

8 If you see a nimbus cloud in the sky, what does it tell you?

____

9 Which two cloud types may be seen in fair weather?

____

10 Which cloud type might you see on either a rainy day or a sunny day?

____

11 What type of cloud is fog?

____

12 Suppose you are on a plane and the pilot tells you that very low clouds make it hard to see for a landing. What type of cloud is most likely to be over the airport?

____

Directions (13–18): Each question is followed by four choices. Decide which choice is the best answer. Circle the letter of the answer you have chosen.

13 If you know it is snowing outside, you know that

A the wind is strong
B the air is cold
C the air is calm
D the clouds are white

14 If it is raining, you know that

A the wind is strong
B the air is warm
C the air is very cold
D the sky is cloudy

15 Rain, snow, sleet, and hail are called

A clouds
B weather
C precipitation
D temperature

16 The temperature tells you

A how warm or cold the air is
B what the weather is like
C whether it is raining or snowing
D how windy the air is

17 The temperature is affected by

A sunlight
B wind
C neither
D both

18 What type of sky might tell you precipitation is probably coming soon?

NYS Test Tip

Earth Science The type of cloud can tell weather reporters whether or not precipitation is on the way.

A clear and sunny
B partly cloudy
C full of thick, dark clouds
D full of white, fluffy clouds

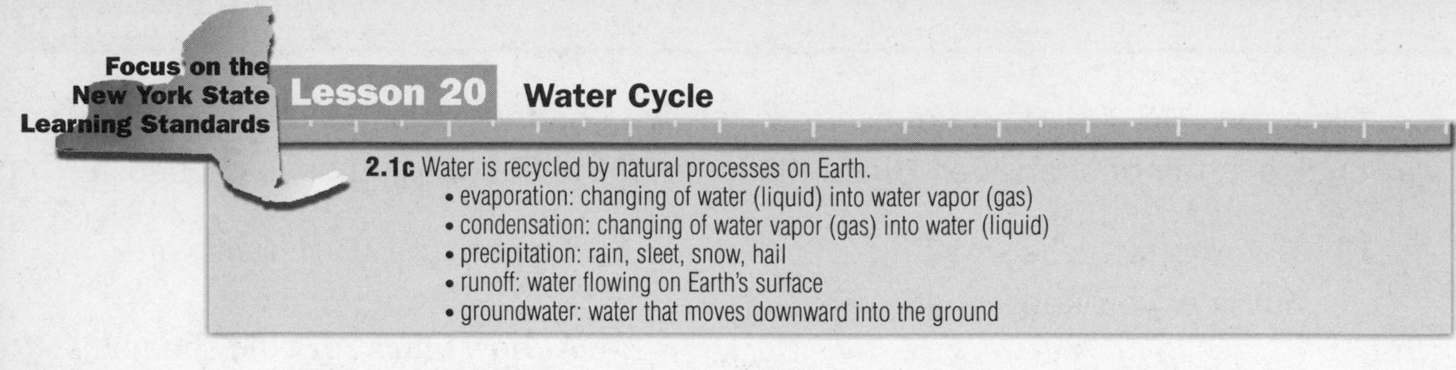

Focus on the New York State Learning Standards

Lesson 20 Water Cycle

2.1c Water is recycled by natural processes on Earth.
- evaporation: changing of water (liquid) into water vapor (gas)
- condensation: changing of water vapor (gas) into water (liquid)
- precipitation: rain, sleet, snow, hail
- runoff: water flowing on Earth's surface
- groundwater: water that moves downward into the ground

Through natural processes, water moves constantly from the ground, to the sky, into the ground, along the ground, and back into the sky.

Water vapor is water in its gas form.

Evaporation is the changing of water (liquid) into water vapor (gas).

Condensation is the changing of water vapor (gas) into water (liquid).

Runoff is water flowing on Earth's surface.

Groundwater is water that moves downward into the ground.

Guided Instruction

Directions Read the following information.

Guided Questions

What would happen if all the water on Earth stayed in the oceans? How would the land get water to grow plants? What would land animals drink? Fortunately, water moves around. Precipitation that falls to Earth is not new water. The same water is constantly recycled through the Earth and the air. The Sun's energy powers what is called the water cycle.

The Water Cycle

Water exists in three states: gas, liquid, and solid. The Sun heats liquid water on Earth, causing **evaporation**, turning it into invisible **water vapor**. Water vapor rises high above the Earth where temperatures are lower. There, **condensation** takes place. The lower temperatures turn the water vapor back into tiny drops of liquid water. These tiny drops form clouds. Rain, snow, hail, or sleet fall to Earth from the clouds. Because wind moves clouds and storms around, the water often falls in a different place from where it evaporated. If the water vapor freezes, it falls as snow. If rain freezes on the way down, it falls as sleet. When pieces of ice form in the clouds from rain or snow, they fall as hail.

Once water lands on Earth, it does not stay in one place. Much of it evaporates right away from the ground or puddles. The rest of the water soaks into the ground as **groundwater**, or runs along the ground as **runoff**. Runoff carries water to rivers, which flow many miles and empty into the oceans. Groundwater also moves slowly toward the rivers and oceans. Water is stored for a time in the ground and in rivers, lakes, oceans, and as ice. Then the heat of the Sun causes it to evaporate, and the cycle starts again.

Because of the water cycle, the land is watered and plants can grow. Water moves from place to place, so people and animals can use water from springs, streams, rivers, and lakes. Humans can dig wells and use the groundwater too.

Guided Questions

How are **water vapor** and **evaporation** related?

What is **condensation?**

What is the difference between **groundwater** and **runoff?**

Directions For each question, write your answer in the space provided.

1. How are groundwater and runoff alike?

____

____

2. What causes condensation of water vapor?

____

3. How are snow and sleet different?

____

____

4. What causes evaporation of water?

____

5. If evaporation is happening all the time, why don't the oceans dry up?

____

____

6. What would happen to Earth's water if wind did not move air, clouds, and storms?

____

____

Apply the New York State Learning Standards to the State Test

Directions: For each question, write your answer in the space provided. Base your answers to questions 7 through 12 on the drawing below.

The Water Cycle

7 What part of the water cycle is shown by the arrows pointing up?

____

8 What part of the water cycle is taking place in the clouds?

____

9 What movement is shown by the arrows running on Earth's surface?

____

10 What movement is shown by the arrows under Earth's surface?

____

11 Where will the groundwater and runoff end up?

____

12 What might explain the fact that water is rising in one area and falling in another?

____

Directions (13–18): Each question is followed by four choices. Decide which choice is the best answer. Circle the letter of the answer you have chosen.

13 What is the constant movement of water from the ground, to the sky, and back again called?

A rain
B snow
C water cycle
D rain clouds

14 Water falling from the sky as rain or sleet is called

A water cycle
B precipitation
C snow
D hail

15 Heavy rain and runoff can

A cause erosion
B cause flooding
C fill reservoirs
D all of the above

16 High clouds are made of

A water vapor
B tiny bits of frozen water vapor
C tiny drops of liquid water
D all of the above

17 What must happen before condensation can take place?

A Water vapor must be high in the sky.
B The temperature must rise.
C Precipitation must take place.
D The Sun must heat the water.

18 A few land areas are lower than sea level. What would you expect to happen to rain that falls on land below sea level?

A It would run off to the ocean.
B It would run off to a river.
C It would soak into the ground.
D It would form a new river.

Focus on the New York State Learning Standards

Lesson 21 Erosion

2.1d Erosion and deposition result from the interaction among air, water, and land.
- interaction between air and water breaks down earth materials
- pieces of earth material may be moved by air, water, wind, and gravity
- pieces of earth material will settle or deposit on land or in the water in different places
- soil is composed of broken-down pieces of living and nonliving earth material

Erosion and deposition move, break down, and build earth materials and structures.

Erosion is the moving and breakdown of earth materials.

Weathering is the breaking of rock into smaller rocks and soil.

Deposition is the dropping or settling of earth materials.

Directions Read the following information.

During rain, you may have watched runoff water collect in puddles on the sidewalk. Often the water in puddles is muddy. The rainwater picks up soil and carries it to the puddle. After the rain stops, the water evaporates, but the dirt carried into it is left behind.

Think about what happens to the ground in the rain. Water carries soil from one place to another. The same process takes place on a larger scale. By a process called **erosion**, water and wind move earth materials around and change the shape of the land they touch. Even gravity causes erosion of land and earth materials when earth and rocks fall.

Part of the erosion process is moving rock and soil from one place to another. Water, wind, air, and gravity also erode land by **weathering**, or breaking rock into smaller rocks and soil. The smaller the rocks or soil pieces, the more likely they are to be moved. Breaking up rocks is not easy. Gases from the air help. They dissolve in water, changing the water so that it can dissolve minerals. After some minerals

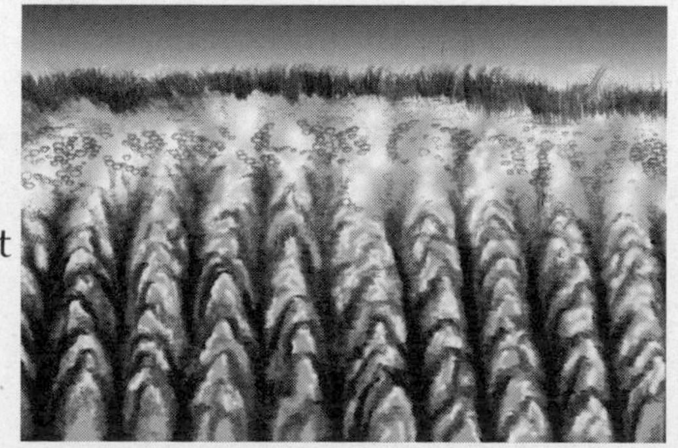

Guided Questions

What is **erosion?**

What is **weathering?**

How has this field been eroded by heavy rain?

in a rock are dissolved, the rock breaks up more easily.

Runoff water moving downhill flows fast. Fast-flowing rivers may move even large rocks. Driven by rushing water, rocks bash into each other. Small chips of rock break off. Over time, water grinds rocks into sand. Sand wears into tiny pieces called silt. Soil is made up of sand, silt, and other broken-down pieces of earth material. Some material in soil was broken down from the bodies of living things.

When moving water and wind slow down, they drop the pieces of sand, silt, and earth they are carrying. After this **deposition**, these tiny pieces of earth form soil miles from where they were picked up or broken down.

After forming, soil can erode and be deposited elsewhere. Carried by a fast-flowing river, rocks and sand scrape and batter the land. Rushing water washes the finest soil from mountains into rivers. Meanwhile, wind picks up dry soil from the land and carries it for miles.

Guided Questions

What happens during **deposition?**

Directions For each question, write your answer in the space provided.

1. What happens during erosion?

____

2. What is weathering?

____

3. How do gases from the air help break down rocks?

____

____

4. How does erosion make more soil?

____

5. Explain how erosion can be helpful to the environment.

____

____

6. Explain how erosion can be harmful to the environment.

____

____

Apply the New York State Learning Standards to the State Test

Directions: For each question, write your answer in the space provided. Base your answers to questions 7 through 12 on the drawing below.

7 When does deposition begin to take place?

____

8 Where is most silt and soil deposited?

____

9 Why does more erosion take place in the mountains?

____

____

10 At which point is more silt in the water: in the mountains or on flatland? Explain.

____

____

____

11 Would you expect more fertile land near the river, in the mountains, or on flat land?

____

12 Where would you expect to find the most good, fine soil in this picture?

____

____

Directions (13–18): Each question is followed by four choices. Decide which choice is the best answer. Circle the letter of the answer you have chosen.

13 Breaking rock into smaller rocks and soil is called

A deposition
B silt
C weathering
D erosion

14 All of the following can cause erosion except

A sunlight
B gravity
C wind
D water

15 What is made up of broken-down pieces of living and nonliving earth material?

A silt
B soil
C sand
D gravel

16 During a rain storm, runoff water collects in puddles on your clean cement driveway. The water in the puddles is muddy. After the rain stops, the water evaporates and you find dried soil where the puddles were. What processes have taken place in your yard?

A weathering and erosion
B weathering and deposition
C erosion and deposition
D weathering, erosion, and deposition

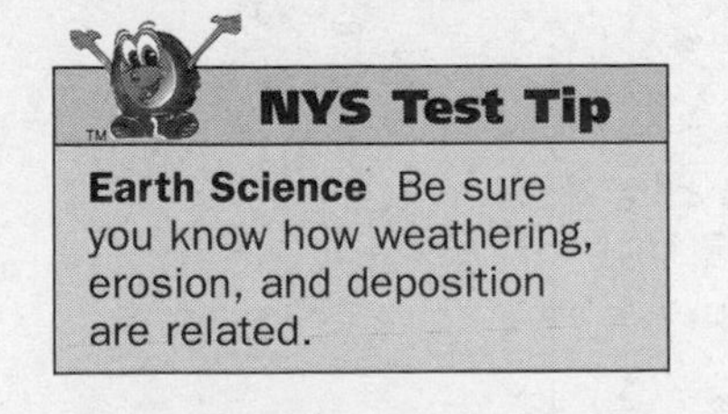

Earth Science Be sure you know how weathering, erosion, and deposition are related.

17 What processes result in a buildup of soil on river banks?

A weathering and erosion
B weathering and deposition
C erosion and deposition
D weathering, erosion, and deposition

18 As moving water slows down, what would you expect it to drop to the bottom of the river first?

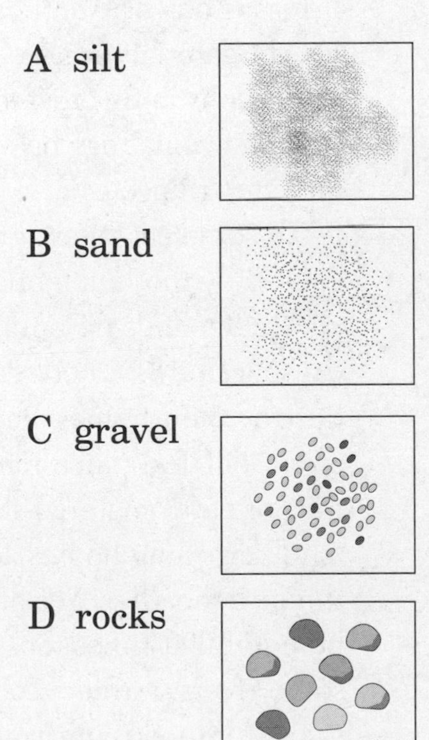

Lesson 22 Extreme Natural Events

2.1e Extreme natural events (floods, fires, earthquakes, volcanic eruptions, hurricanes, tornadoes, and other severe storms) may have positive or negative impacts on living things.

Natural earth processes can create extreme natural events.

A **flood** is an overflowing body of water.

A **tornado** is a cloud shaped like a funnel that spins.

A **hurricane** is a tropical storm with strong winds.

An **earthquake** is a sudden, strong movement of Earth's crust.

A **volcano** is an opening in Earth's crust from which underground steam, ash, gas, and hot liquid rock escape.

Directions Read the following information.

Natural processes take place all the time on Earth. The water cycle sends water to the sky and brings it down. Weather changes from day to day. Earth materials are eroded and deposited. New soil is constantly created. Sometimes, though, natural events may be extreme with major impacts on living things.

In some cases, extreme natural events are caused by too much of a good thing. For example, all life depends on rain, but too much rain can cause a **flood**. Floods can drown living things, destroy their homes, and wash away soil. Too much rain can bring tons of earth material sliding down a hill, crushing homes and living things beneath it. Yet floods can be helpful at times too. Along some rivers, farmers count on the rivers to flood their banks and the surrounding valleys. The floods water the land and deposit silt for farming.

Tornado

Guided Questions

What is a **flood?**

The water cycle, temperature, and wind combine to make storms and precipitation. Life depends on these storms, but some are fierce and dangerous. Sometimes a thunderstorm will cause a **tornado** to form. A whirling funnel of air extends down from the thundercloud. Winds spin in a tight circle at terrific speeds. When a tornado whirls through a town, it may destroy everything it touches.

Storms called **hurricanes** can travel hundreds or thousands of miles. Their winds are less violent than those of tornadoes, but hurricanes may be hundreds of miles wide and last for several days. Terrific winds uproot trees and level buildings. Heavy rainfall may wash whole neighborhoods away. Many living things are destroyed.

Extreme events can also result from natural processes in Earth. Underground movements can result in **earthquakes**. Violent shaking topples buildings and bridges. Hot lava, steam, and ash from underground can erupt from **volcanoes**. The lava and ash can burn or bury living things near the volcano. Yet some of Earth's most beautiful mountains and islands have been created by erupting volcanoes. And, the soil formed from lava is rich in nutrients for crops.

Guided Questions

What is a **tornado?**

What is a **hurricane?**

What is an **earthquake?**

What is a **volcano?**

Directions For each question, write your answer in the space provided.

1. What is a tornado and when may it form?

____

____

2. What makes a hurricane dangerous?

____

3. Give an example of a positive and a negative effect of floods.

____

____

____

____

4. What causes an earthquake and what negative effect can it have on the environment?

____

____

____

5. If a volcano erupts many times, a mountain sometimes builds up around it. Why do you think this happens?

____

6. Explain how too much rain or the lack of rain can cause extreme problems for a farmer.

____

____

____

Apply the New York State Learning Standards to the State Test

Directions: For each question, write your answer in the space provided. Base your answers to questions 7 through 12 on the drawing below.

Volcano

7 What do you call the opening in these mountains?

____

8 What is happening in the underground tunnels?

____

9 How do you think this cone-shaped mountain was built?

____

10 What will happen to the land and living things in the path of the lava flow?

____

11 What will happen to the lava that has erupted from the volcano?

____

12 Think about the steam and hot lava that erupt from a volcano. What do they tell about conditions deep in Earth's core?

___

Directions (13–18): Each question is followed by four choices. Decide which choice is the best answer. Circle the letter of the answer you have chosen.

13 Which is not an extreme natural event?

A flood
B earthquake
C hurricane
D rain

14 Hurricanes are caused by

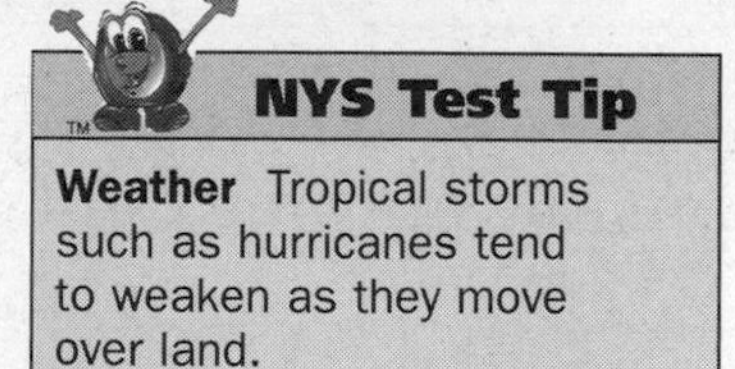

A the same natural processes that create ordinary weather
B rare processes that rarely occur
C the joining of several large tornadoes
D volcanic eruptions that affect the condition of the air

15 Floods affect living things in ways that are

A positive
B negative
C positive or negative
D neither positive nor negative

16 Which statement about tornadoes and hurricanes is true?

A Tornadoes are more destructive than hurricanes.
B Hurricanes have faster wind speeds than tornadoes.
C Tornadoes are more violent but hurricanes damage a larger area.
D Tornadoes and hurricanes cause an equal amount of damage.

17 If you see a whirling funnel of air descending from a dark cloud, you know that

A a hurricane is forming
B a tornado is near
C a volcano has erupted
D a volcano is erupting

18 Which two events are most likely to happen at the same time in the same place?

A hurricane and flooding
B flooding and volcano erupting
C volcano erupting and tornado
D tornado and hurricane

Focus on the New York State Learning Standards

Lesson 23 Earth, the Moon, and the Sun

1.1a Natural cycles and patterns include: Earth spinning around once every 24 hours (rotation), resulting in day and night; Earth moving in a path around the Sun (revolution), resulting in one Earth year; the length of daylight and darkness varying with the seasons; weather changing from day to day and through the seasons; the appearance of the Moon changing as it moves in a path around Earth to complete a single cycle.
1.1b Humans organize time into units based on natural motions of Earth: second, minute, hour, week, month.
1.1c The Sun and other stars appear to move in a recognizable pattern both daily and seasonally.

The regular pattern of movement by Earth, the Moon, and the Sun affects cycles and conditions on Earth's surface.

Rotation is Earth spinning around once every twenty-four hours, resulting in day and night.

Revolution is Earth moving in a path around the Sun, resulting in one Earth year.

Directions Read the following information.

Many natural cycles and patterns on Earth's surface result from natural cycles and patterns in the movement of Earth, the Moon, and the Sun. The movements of these bodies are so important that human calendars and clocks are based on them.

An important Earth cycle that affects living things every day is its **rotation**. Once every twenty-four hours, Earth spins around on its axis, an imaginary line running through its center from one pole to the other. At any given moment, half of Earth's surface is in sunlight and half is in darkness. Because different parts of it face the Sun at different parts of the rotation, this cycle results in day and night on Earth's surface.

Guided Questions

What is **rotation?**

People divide the day into twenty-four hours. Each hour is divided into sixty minutes, and each minute into sixty seconds. A week is seven days.

All life on Earth is also affected by its **revolution**. Earth moves in a path around the Sun. One turn around the Sun is one revolution. One Earth year is based on the length of time it takes to revolve around the Sun once. Because Earth is tilted on its axis, the direct rays of the Sun are focused on different places on Earth at different times of the year. This causes seasons. The length of daylight and darkness varies with the seasons. As the amount of sunlight changes, weather changes from day to day and through the seasons.

As the Moon moves in a path around Earth, different parts of it are lit by the Sun at night. Therefore its appearance changes. The different ways it appears are called phases of the Moon. What we call a month is about the length of time it takes the Moon to complete a single cycle around Earth.

When humans watch the Sun and other stars from Earth, they appear to move in a pattern. Every day, the Sun comes up in the east and sets in the west. The same stars appear in the same patterns at times scientists can predict.

Guided Questions

What is **revolution?**

Directions For each question, write your answer in the space provided.

1. About how long does it take the Moon to circle Earth?

___

2. How did humans decide how long a year should be?

___

3. How is life on Earth affected by Earth's revolution around the Sun?

___

___

4. Why does the appearance of the Moon change throughout its cycle?

____

5. Suppose Earth did not rotate. How would conditions and life on Earth be different?

____

____

____

____

Apply the New York State Learning Standards to the State Test

Directions: For each question, write your answer in the space provided. Base your answers to questions 6 through 11 on the drawing below.

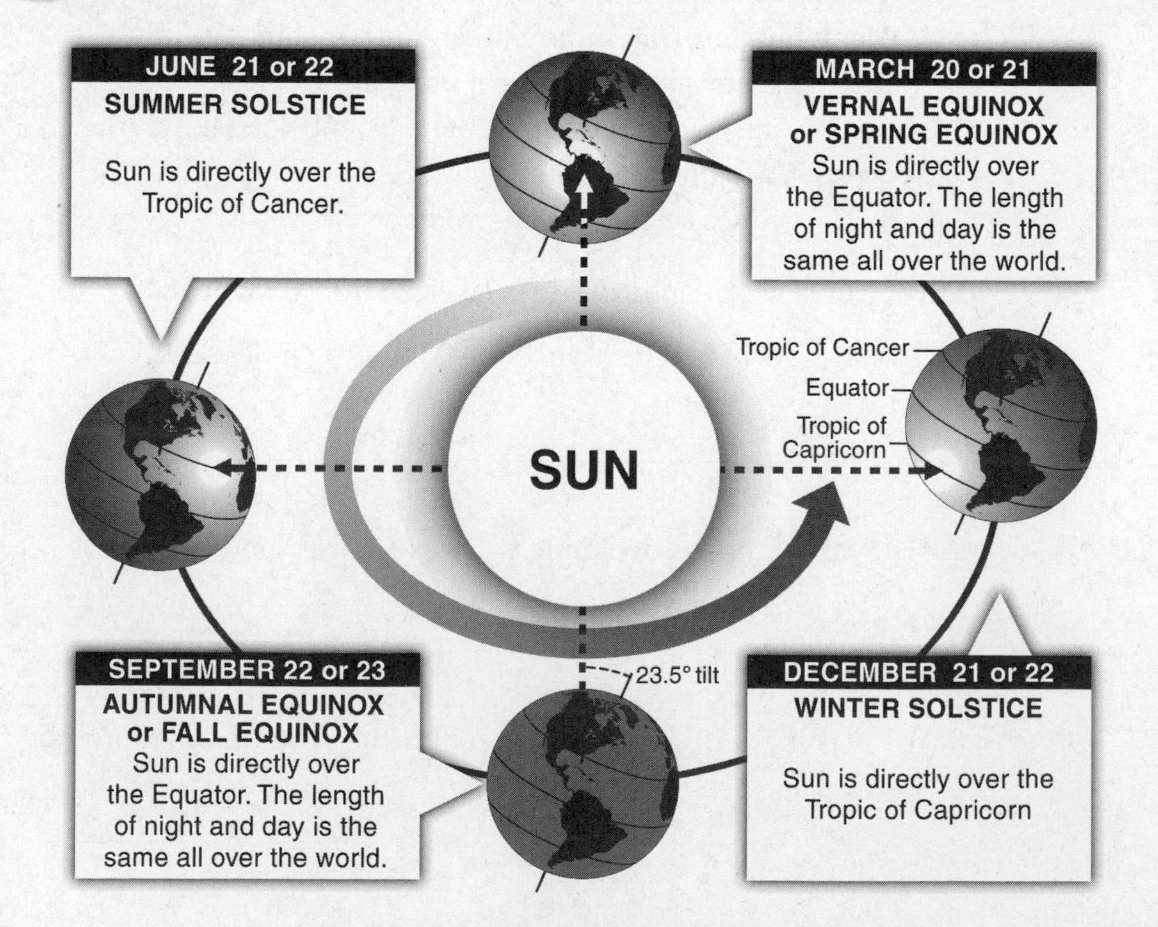

6 What is the line running through Earth from pole to pole?

____

7 In the United States, when is the length of day longest?

____

8 What do you notice about Earth's axis during the winter solstice?

____

9 Why is the length of day the same all over the world at the spring equinox?

____

____

10 What part of Earth is darkest at the solstice? Why?

____

____

11 At what time of year would the tip of South America be warmest?

____

Directions (12–17): Each question is followed by four choices. Decide which choice is the best answer. Circle the letter of the answer you have chosen.

12 What causes night and day on Earth?

A rotation of Earth
B revolution of Earth
C rotation of the Moon
D revolution of the Moon

13 What determines the length of night on Earth?

A rotation of Earth
B rotation of the Moon
C both
D neither

14 A week is

A the length of time it takes the Moon to revolve around Earth
B the length of time it takes Earth to revolve around the Sun
C both
D neither

15 The rising and setting of the Moon is mostly caused by

A rotation of Earth
B revolution of Earth
C the shadow of Earth
D rotation of the Sun

16 How long does it take the Moon to circle Earth?

A about a week
B about a year
C about a season
D about a month

17 How would conditions on Earth be different if its axis did not tilt?

A There would be no seasons.
B There would be no years.
C There would be no temperature.
D The equator would not be as warm in winter.

Higher-Order Performance Task

Investigating Day and Night

Task:

You will be making a model of Earth and the Sun. Then you will use your model to investigate why some places on Earth have day while others have night.

Materials:

- globe
- flashlight
- sticker or tape
- stopwatch

Directions:

1. Use the materials to complete this task.
2. Find New York on the globe. If a globe is not available, you can sketch a map on an inflated balloon or a ball. Place a sticker or piece of tape on the globe to show where New York is.
3. Turn on the flashlight. Lay it flat on a desk or table. Move the globe so that it is lit by the flashlight. Dim the lights in the room.
4. What does the flashlight represent in the model?

   ____

5. Turn the globe so that New York is in the light from the flashlight. How do you know that New York is having daytime?

   ____

6. Name a place on Earth that is having night while New York has day.

   ____

7. Slowly spin the globe in a counterclockwise direction. That means that if you look down at the North Pole, it will go in a direction that is opposite to the direction in which the hands of a clock move. Spin the globe until New York has night. How do you know that New York is having night?

____

____

8. Name a place on Earth that is having day while New York has night.

____

9. The Sun rises in the east and sets in the west. Based on your model, explain why this is true.

____

____

____

10. Use the materials to design an activity to show that the length of day and night depends on how fast Earth spins. Describe the activity below. With your teacher's permission, demonstrate your activity.

____

____

____

11. When you have finished, put all the materials back the way you found them.

Building Stamina®

Directions (1–18): Each question is followed by four choices. Decide which choice is the best answer. Circle the letter of the answer you have chosen.

1 You see clouds that look like pieces of fluffy white cotton. What kind of weather can you expect?

A rain
B fog
C sun
D snow

2 During a hard rain, a farmer's field may become flooded because the extra rainwater cannot soak in. This water is called

A runoff
B groundwater
C evaporation
D precipitation

3 Water from a recent rainstorm has been soaked down into the grass and soil. This water is now considered to be

A runoff
B groundwater
C precipitation
D evaporated

4 Which of the following is *not* an extreme natural event?

A earthquake
B hurricane
C snowfall
D tornado

5 Weather can be described and measured by

A temperature
B wind speed and direction
C kind of precipitation and clouds
D all of the above

6 Sand storms occur in places such as a desert. Which kind of erosion may occur on the rocks and hills in the desert?

A water erosion
B wind erosion
C glacial erosion
D erosion due to gravity

7 Which list below correctly describes what soil for growing crops is made up of?

A weathered pieces of rock
B weathered pieces of rock and water
C weathered pieces of rock, minerals, water, and air
D weathered pieces of rock, dead plants and bugs, minerals, water, and air

8 Eroded material that gets deposited in lake bottoms and rivers is called

A earth material
B rock fragments
C soil
D sediment

9 You see a sign along the highway that says "Falling Rocks." What is most likely the cause of the falling rocks?

A wind erosion
B water erosion
C vibration from the cars
D gravity

Base your answers to questions 10 and 11 on the drawing below.

10 Which part of the drawing above shows precipitation?

A 1
B 2
C 3
D 4

11 Which part of the drawing above shows where condensation takes place?

A 1
B 2
C 3
D 4

12 Hurricanes often cause salt water to flood into environments that have fresh water. When this happens, the plants living in the fresh water would most likely

A grow more rapidly
B die
C shrivel up
D fill up with more water

13 Geologists study earthquakes to try to find a way to protect people. Which of the following is *not* a positive result of their research?

A Don't build homes on fault lines.
B Design earthquake-proof buildings.
C Have emergency drills in schools in areas where earthquakes are common.
D Build roadways along fault lines to avoid getting cracks in them when an earthquake occurs.

14 The Moon changes in how it appears because

A Earth moves around the Moon
B The Moon moves around Earth
C The Moon moves around the Sun
D Earth moves around the Sun

15 The Earth has seasons because

A Earth rotates
B the Moon revolves around Earth
C Earth is tilted on an axis and it revolves around the Sun
D Earth is closer to the Sun at different times during the year

16 The drawing below shows how Earth moved from the letter A, around the Sun and back to the letter A. How much time has gone by?

A 1 hour
B 1 day
C 1 month
D 1 year

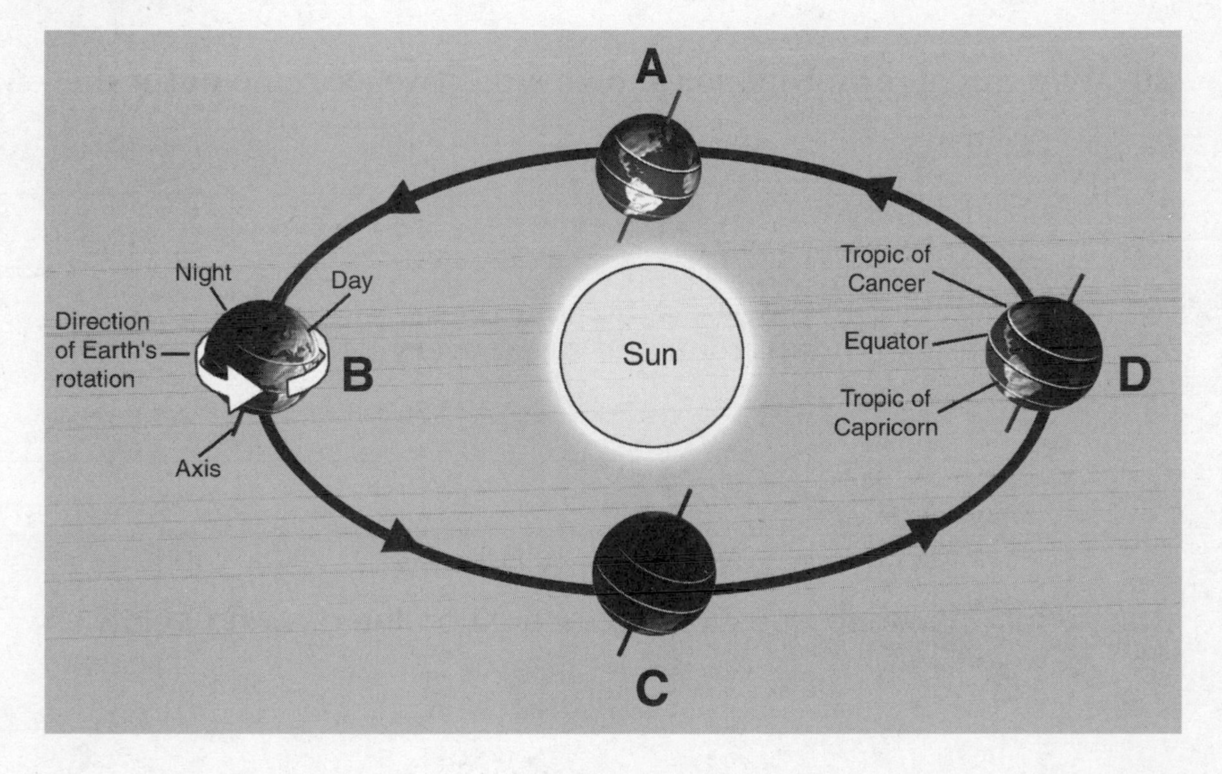

17 Seven Earth rotations is the same amount of time as one

A day
B week
C month
D year

18 About how much time has gone by after the Moon circles Earth once?

A 1 hour
B 1 day
C 1 month
D 1 year

Directions: For each question, write your answer in the space provided.

19 The temperature outside is 28°F. Explain what type of precipitation will most likely fall at this temperature.

____

____

20 Why can an erupting volcano be a positive extreme natural event?

____

____

21 Why does Earth have day and night every 24 hours?

____

____

Base your answer to questions 22 and 23 on the drawing below.

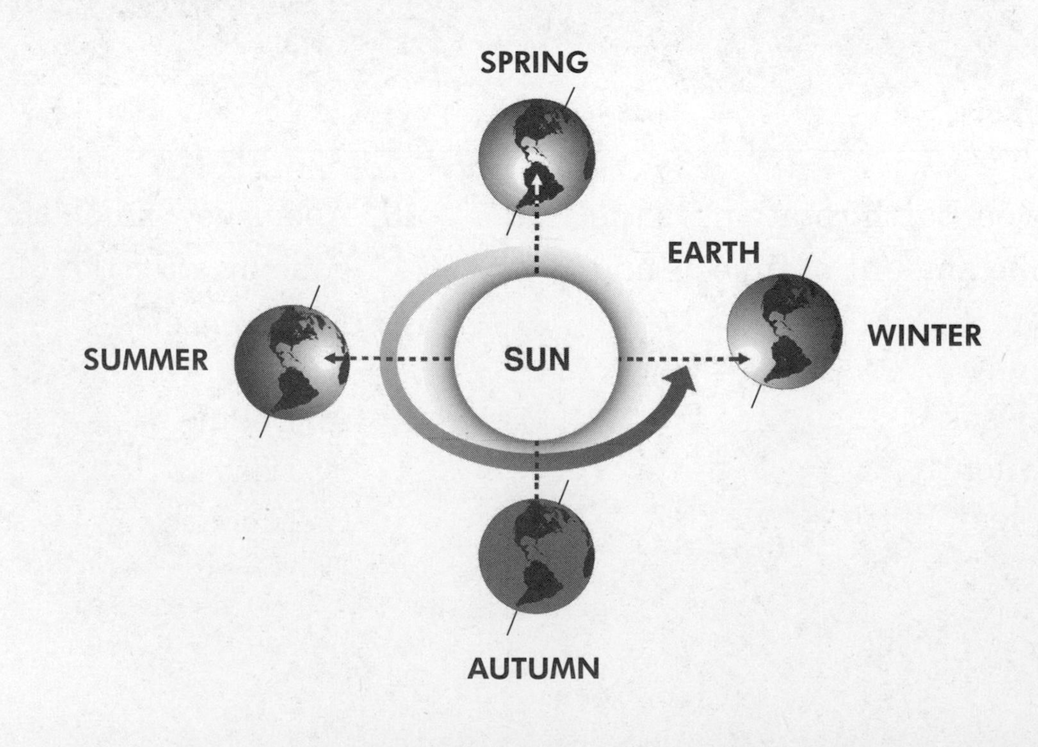

22 Explain why Earth has seasons.

____

____

____

23 Why is the constellation Orion, the hunter, visible to North America only in the winter months?

____

____

____

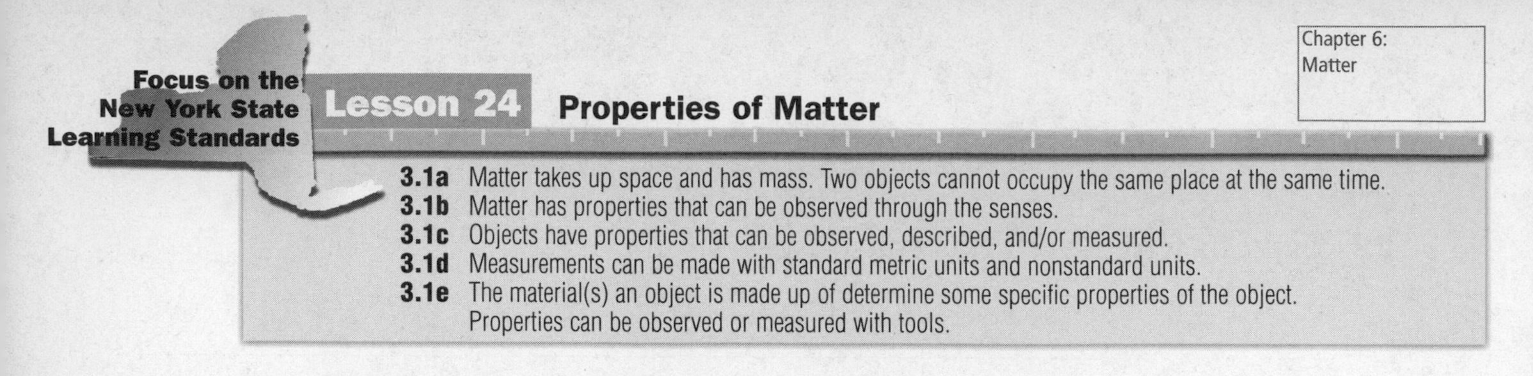

Properties of Matter

3.1a Matter takes up space and has mass. Two objects cannot occupy the same place at the same time.
3.1b Matter has properties that can be observed through the senses.
3.1c Objects have properties that can be observed, described, and/or measured.
3.1d Measurements can be made with standard metric units and nonstandard units.
3.1e The material(s) an object is made up of determine some specific properties of the object. Properties can be observed or measured with tools.

Using your senses and appropriate tools, you can observe, measure, and describe the properties of matter.

A **property** is what can be observed about an object. Size, shape, color, hardness, taste, and weight are properties of an object.

Matter is anything that takes up space and has mass. Matter is made up of particles that have properties that can be observed through our senses.

Mass is the amount of matter an object contains. Mass is measured in grams.

Guided Instruction

Directions Read the following information.

Matter is all things around you. It is anything that takes up space or has **mass**. Everything you can touch is made of matter, and no two objects made up of matter can be in the exact same place at the same time. Matter has many **properties** that we can observe by using our senses of sight, touch, hearing, smell, and taste. Think about the orange you might have for lunch.

- It looks round or spherical and has an orange color.
- When you touch it, it has a slightly bumpy texture and might feel soft.
- When you peel it, you might hear a crisp, ripping sound.
- When you smell it, it smells like an orange.
- When you taste it, it will taste sweet if you have a good one!

orange

Guided Questions

What **properties** of an orange can you observe through your senses?

The materials that an object is made up of determine some of its properties. We know that a cork, which comes from a tree, will float in water, but how about a metal iron nail? You're right if you said it will sink. Another property of an iron nail is its attraction to a magnet, but not all metals have this magnetic property. Neither a copper penny nor a piece of aluminum foil will be attracted to a magnet. Metal objects with iron in them are attracted to a magnet. Many metal objects are made of steel, which has iron in it.

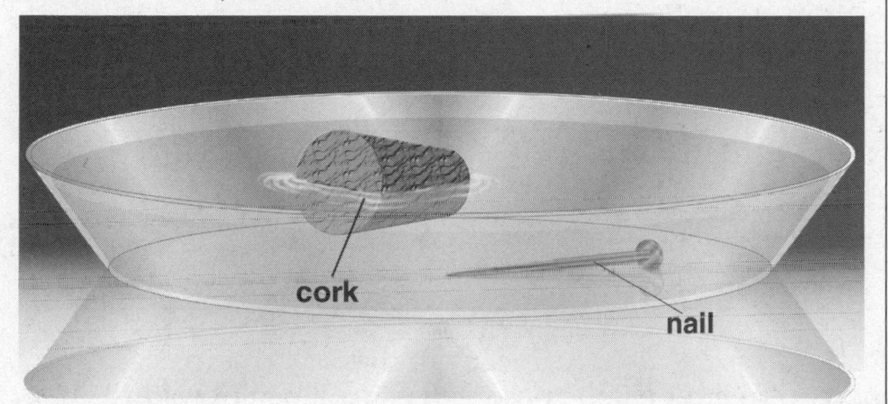

Copper has a property that it does not share with aluminum. Copper is a good conductor of electricity. Aluminum is a poor conductor of electricity. That is why copper is used in the wiring in electrical appliances and in the wires in your home. Good or poor conductivity is another property of matter.

Some metals conduct something else. They conduct heat. That is why you should never touch the handles or sides of a hot cooking pot or pan unless you use a protective hot pad or oven mitt.

Length, width, volume, size, shape, mass or weight, and temperature are also properties that help us describe an object.

Guided Questions

How can you test for the property of magnetism?

How are the properties of cork and iron different?

Are the properties of all metals the same? Give an example of your answer.

Directions For each question, write your answer in the space provided.

1. Where is matter found?

____

2. Use your five senses to describe hot buttered popcorn.

____

____

____

____

____

3. What property of matter can be observed using a pan of water?

____

4. Paper clips are made of steel, which has iron in it. Do you expect paper clips to be attracted or not attracted to a magnet? Why or why not?

____

____

5. What properties of an object would you observe when you use a metric ruler?

____

6. What metric unit is used to describe the mass of an object?

____

Apply the New York State Learning Standards to the State Test

Directions: For each question, write your answer in the spaces provided. Base your answers to questions 7 through 12 on the paragraph below.

Your class is studying the properties of matter. Your group has been given a rock sample to examine. First you notice that the rock is pinkish-gray in color and that it has a rough texture. You use a hand lens to take a closer look and notice tiny particles. Your group uses a balance and finds that the mass of the rock sample is 90 grams. Using a metric ruler, you find that the length of the sample is 2 centimeters. Your group thinks that this rock sample will sink and not float. To check out this prediction, you place the sample in a pan of water. Your prediction was right. It sinks.

color	pinkish-gray
texture	**7** ________________
mass	**8** ________________
9 ________________	2 cm
sink or float?	**10** ________________

11 What tool would help the students get a close look at the rock sample?

____

12 What tool would help the students find the mass of the rock sample?

____

Directions (13–18): Each question is followed by four choices. Decide which choice is the *best* answer. Circle the letter of the answer you have chosen.

13 Why should you never pick up a hot pot by its metal handle?

A The handle will be hot because metal is a good conductor of heat.
B The handle will be slippery because metal is slippery when heated.
C The handle will be soft because the metal in cooking pans is easily melted.
D The handle will be freezing cold because metal turns heat to cold.

14 Which property of an object will never affect whether the object sinks or floats?

A shape
B weight
C color
D width

15 What is the length to the nearest centimeter of the paper clip shown below?

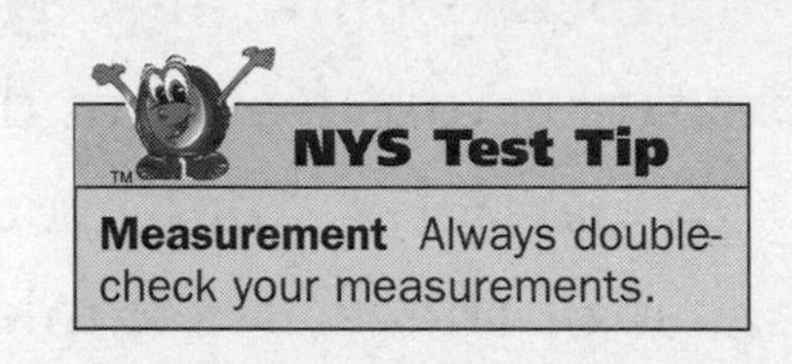

A about 3 cm
B about 4 cm
C about 5 cm
D about 6 cm

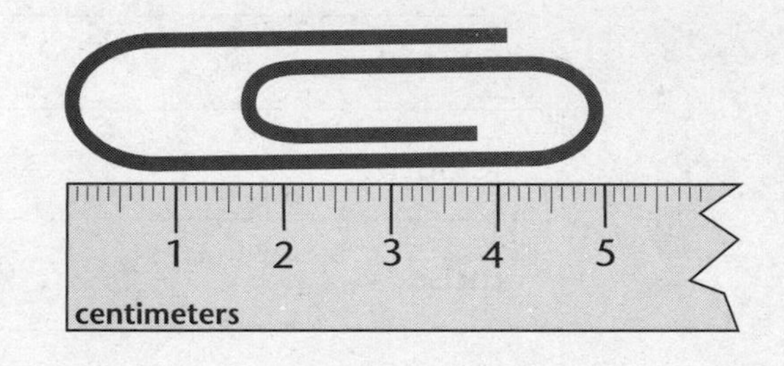

16 You notice that paper clips are attracted to a box. Which statement best explains why this may be happening?

A The paper clips are lighter than the box.
B The box is made of cardboard.
C The paper clips are hard
D A magnet is inside the box.

17 A student found a rock while hiking in the mountains. By looking at the rock, she could tell

A the exact weight of the rock
B the length of time the rock had been on the hiking path
C the color and shape of the rock
D the exact length of the rock

18 The chart below lists some objects and shows if a bar magnet picks them up. Choose the conclusion that can be drawn from the information in the chart.

OBJECT	IS IT ATTRACTED TO THE MAGNET?
staple	yes
eraser	no
iron nail	yes
a penny	no
pin	yes

A The magnet attracted none of the objects.
B The magnet attracted all the metal objects.
C The metal objects with iron in them share the property of magnetism.
D The rubber objects share the property of magnetism.

Focus on the New York State Learning Standards

Lesson 25 Classifying Matter

3.1e The material(s) an object is made up of determine some specific properties of the object. Properties can be observed or measured with tools.
3.1f Objects and/or materials can be sorted or classified according to their properties.
3.1g Some properties of an object are dependent on the conditions of the present surroundings in which the object exists.

You can observe and measure objects, and you can classify objects according to their properties.

Classify means to sort or arrange objects into groups based on how they are alike or how they are different.

Properties are what can be observed about an object. Size, shape, color, hardness, taste, texture and weight are properties of an object.

Guided Instruction

Directions Read the following information.

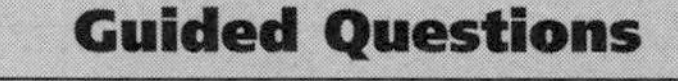

Whenever you sort or organize objects, you are **classifying** them according to one or more **properties**. If you have a handful of change, you might sort the coins by their value. All the pennies would go into one group and all the dimes would go into another group. How else might you sort the coins? You might use color. All the silver coins would go into one group and the copper-colored coins would go into a second group as shown below.

What other **properties** could you use to **classify** coins?

Objects, which are made up of matter, can be sorted or classified according to their properties. Properties can be observed or measured with tools such as hand lenses, metric rulers, thermometers, balances, magnets, circuit testers, and graduated cylinders. The material or materials that make up an object can give it other properties too. An object might sink or float. It might or might not conduct heat or electricity. It might or might not be attracted to a magnet.

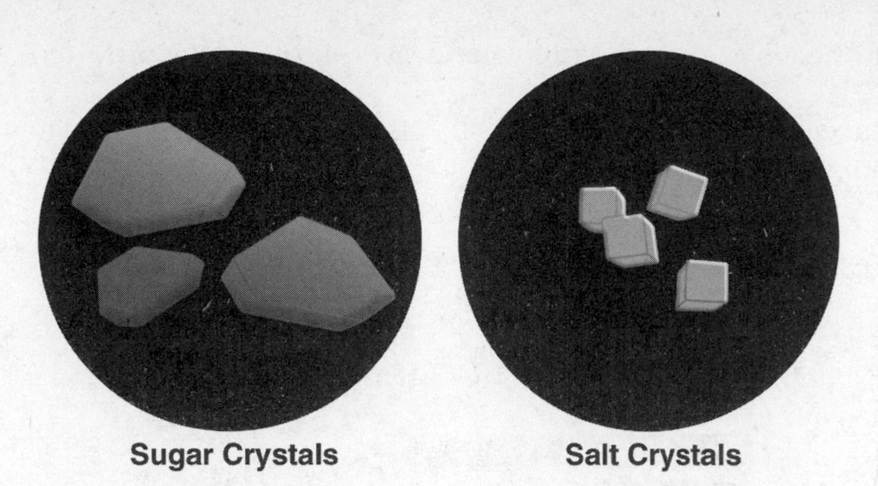

Sugar Crystals Salt Crystals

Let's look at the properties of two things you might find in a kitchen: brown sugar and salt. Sugar and salt are crystals. But the shape of sugar crystals and the shape of salt crystals are different. You can use a hand lens to see this difference.

If you had a spoonful of salt and a spoonful of brown sugar, you could classify them in different ways. One way would be to classify both materials into a single group called crystals. Another way would be to classify them by shape of crystal. Then you would have two groups, sugar-shaped crystals and salt-shaped crystals. Could you use the property of color to classify the brown sugar and salt? If you did, you would have a brown group and a white group.

One of the properties of water depends upon the temperature of the water. Warm water is liquid and takes the shape of the container it is in. Frozen water, or ice, is a solid and it holds its shape. If you had two glasses of water and two ice cubes, you could classify them in different ways. One way would be to put the glasses and the ice cubes into one group because they are all water. Another way would be to classify them into two groups: the glasses of water would be in a group of liquids; the ice cubes in a group of solids.

Guided Questions

How are the shapes of the crystals different?

What other property could you use to classify sugar and salt?

Directions For each question, write your answer in the space provided.

1. What do we call sorting or arranging objects into groups based on how they are alike or how they are different?

____

2. What tool is being used to measure these two objects?

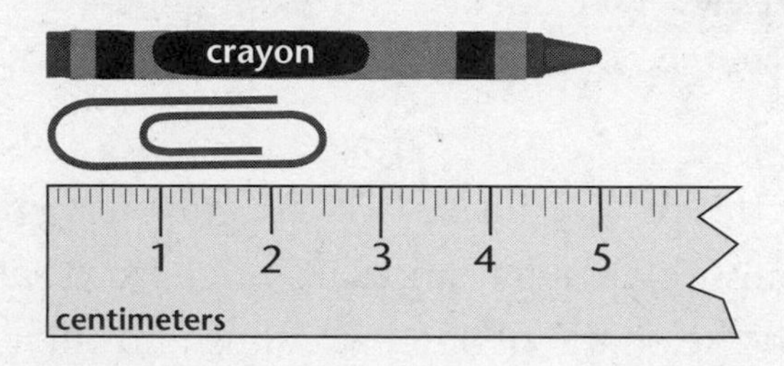

____

3. What is the name of this tool What does it help you do?

____

____

4. You have a box of buttons. What properties might you use to classify them?

____

____

5. How is water affected by temperature?

____

6. If you tinted a glass of warm water with blue food coloring, what property or properties of the water would be different?

____

Apply the New York State Learning Standards to the State Test

Directions: For each question, write your answer in the spaces provided. Base your answers to questions 7 through 9 on the paragraph and chart below.

Your science class has just observed several rock samples. The students recorded their observations in the following chart. Your job is to use the observations to classify, or sort, the rock samples.

ROCK	OBSERVATION
slate	black; harder than shale
shale	a darkish color; smells musty when it is wet
sandstone	grainy; light brown to pinkish in color
pumice	floats in water; light-colored
obsidian	black; shiny
lodestone	dark; attracted to a magnet

7 Sort the samples into light and dark by writing the name of the rock in the correct column.

LIGHT COLOR	DARK COLOR

8 Sort the samples into rocks that float and rocks that sink by writing the name of the rock in the correct column.

FLOAT	SINK

9 Sort the samples into rocks that are attracted to a magnet and those that are not attracted to a magnet by writing the name of the rock in the correct column.

ATTRACTED TO A MAGNET	NOT ATTRACTED TO A MAGNET

Directions (10–15): Each question is followed by four choices. Decide which choice is the *best* answer. Circle the letter of the answer you have chosen.

10 When you classify objects, you

A paint them
B sort them
C list each object
D give each object a name

11 Size, shape, color, hardness, taste, texture, and weight are

A measured by a metric ruler
B observed through a hand lens
C properties
D determined by the temperature of an object

12 Which two tools would help you measure the size and mass of an object?

A hand lens and balance
B metric ruler and magnet
C metric ruler and balance
D thermometer and hand lens

13 Which properties best describe a U.S. penny (1 cent)?

A soft, gray, smooth
B soft, coppery, cold
C bright, gold, bumpy
D coppery, hard, round

14 A student found six shells on the beach. Three shells were pink, three were cone-shaped, and two were much larger than the other four. Knowing only these properties, what four ways can a student classify these shells?

A by size, by color, by shape, by texture
B as shells, by size, by color, by shape
C by temperature, by color, by sink or float, by shape
D as shells, by weight, by taste, by size

NYS Test Tip

Physical Science You can observe properties of matter by seeing, touching, hearing, smelling, and tasting.

15 Look at the drawings of the four buttons.

What properties would you use to sort the buttons so there are two buttons in each group?

A buttons with stripes and buttons without stripes
B buttons with three holes and buttons with two holes
C buttons that are round and buttons that are square
D buttons with holes and buttons without holes

Focus on the New York State Learning Standards

Lesson 26 Changes of Matter

3.2a Matter exists in three states: solid, liquid, gas.
3.2b Temperature can affect the state of matter of a substance.
3.2c Changes in the properties or materials of objects can be observed and described.
4.1f Heat can be released in many ways, for example, by burning, rubbing (friction), or combining one substance with another.

You can describe chemical and physical changes of matter.

Matter is everything around you that takes up space.

State of matter tells if a substance is a solid, liquid, or gas.

Temperature is the degree of heat or cold of a substance.

Guided Instruction

Directions Read the following information.

Everything around you is **matter**. Your desk, other students, the air, and the water in the drinking fountain are all different kinds of matter. Matter has different forms, called states. The three **states of matter** we can observe are solids, liquids, and gases.

When matter is a solid, it holds its shape. Its volume, which is the space it fills, stays the same. When matter is a liquid, its shape can change, but its volume stays the same. When matter is a gas, it cannot hold its shape or its volume.

Three States of Matter

Guided Questions

What are the three **states of matter?**

When is water a solid?

Water is the only substance that exists naturally as all three states of matter:

- Water is a liquid when it rains or flows from a faucet.
- Water is a solid when it is snow or frozen ice.
- Water is a gas when heat causes it to become invisible water vapor in the air.

Temperature is how hot or cold something is. Water changes into a solid when the temperature of the water decreases and the water freezes into ice. Water in the freezer of your refrigerator turns to ice. You might use ice cubes to cool a drink. Water at the cold North and South Poles of Earth is frozen solid into glaciers. Water changes into gas when the water is heated and the temperature increases. The burner on a stove heats water in a pan and the water turns to water vapor. Water vapor is an invisible gas. The steam you see above a pot of boiling water is water vapor moving into colder air and changing into water droplets. This steam is like a small cloud. When the steam evaporates, it changes into water vapor. When water vapor meets a cold surface, such as a cold window, it turns back into liquid water. The Sun provides the heat that changes rain puddles into water vapor. When the water vapor rises in the air, it cools, and changes into liquid water droplets that you see as clouds.

Other substances can change their state of matter, too. Metal, which is usually a solid, can be heated to turn it to liquid. Some jewelry is formed this way. The metal is heated until it is a liquid. Then it is poured into a mold. When the metal cools and becomes a solid again, it holds the shape of the mold. Rock, which is a solid, might be heated deep below Earth's surface and then erupt from a volcano as molten, or liquid, lava. When the lava cools on Earth's surface, it changes back to a solid.

Guided Questions

What happens to water vapor when the **temperature** decreases?

What is one way to change a liquid to a solid?

How do you change a solid to a liquid?

Directions For each question, write your answer in the space provided.

1. Explain why a wooden block is a solid.

____

____

2. Which of the three states of matter cannot hold a shape or volume?

____

3. If a liquid is poured from a tall bottle into a shallow pan, does its shape or volume change? Explain.

____

____

4. What will most likely happen if a drop of liquid water falls on a hot stove?

____

____

5. What change occurs if solid metal is heated to a very high temperature?

____

____

6. What happens when molten, or liquid, rock cools on Earth's surface?

____

____

Apply the New York State Learning Standards to the State Test

Directions: For each question, write your answer in the spaces provided. Base your answers to questions 7 through 11 on the drawing below.

7 What three states of matter are shown in this diagram?

____

____

8 What change will occur to the ice cubes if they are placed into the boiling water?

____

____

9 Water vapor is an invisible gas. What is steam?

____

____

10 Did you ever notice that ice can sometimes form on the inside walls of a freezer? Where do you think the ice comes from?

____

____

____

11 Compare the properties of solids, liquids, and gases.

____

____

____

Directions (12–17): Each question is followed by four choices. Decide which choice is the *best* answer. Circle the letter of the answer you have chosen.

12 Which of the following holds its shape and volume?

A gases
B liquids
C solids
D oceans

13 In what state of matter would you find rock as it erupts from a volcano?

A gas
B liquid
C solid
D red

14 Which activity would change a liquid to a solid?

A crushing ice
B catching rain in a bucket
C melting snow
D making ice cubes

15 Which place would most likely turn a stick of butter into a liquid?

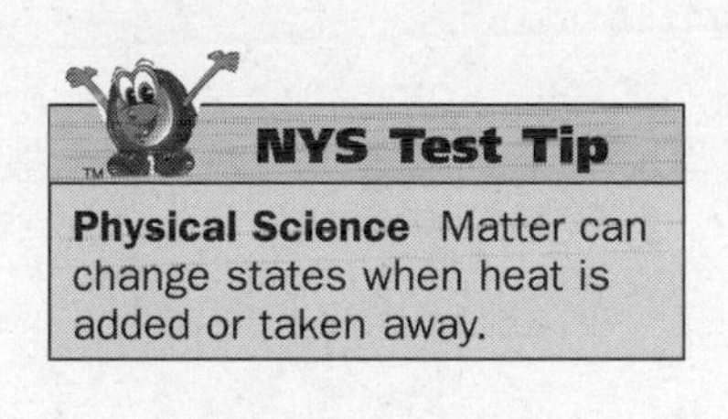

A a freezer
B a cool room
C a hot stove
D a refrigerator shelf

16 If the world's temperatures increased and the glaciers melted, Earth would have

A more gas
B more solid
C more liquid
D more ice

17 A cloud is made of water vapor and condensed water vapor. In what state, or states of matter, is a cloud?

A gas and liquid
B liquid
C solid and gas
D cotton

Higher-Order Performance Task

Change the State of Water

Task:

You will set up a small environment. Then you will make water change from one state to another as it does in a real environment.

Materials:

- large metal bowl
- pitcher or bucket
- clear plastic wrap
- dry ceramic mug
- large rubber band
- water
- ice cube
- paper towels

Directions:

1. Place the bowl in direct sunlight.
2. Use the pitcher or bucket to fill the bowl about ¼ full with water..
3. Place the mug in the center of the bowl. The water level should not be above the top of the mug.
4. Stretch plastic wrap across the top of the bowl.
5. Stretch the rubber band around the top so it holds the plastic wrap in place. You may need to have a partner hold the plastic wrap while you stretch the rubber band.

6. Observe the bowl for several minutes. Write any observations in the first row of the data table below.

	Observations
Before Ice Cube	
After Ice Cube	

7. Gently place an ice cube on top of the plastic wrap. Observe again. Write any observations in the second row of the data table below.

8. What was the purpose of adding the ice to the top of the plastic wrap?

____

9. What happened to some of the liquid water when it was heated by the Sun?

____

10. What formed on the inside of the plastic after you added the ice cube?

____

11. How did the ice cube cause water to change from a gas to a liquid?

____

____

____

12. Carefully pull back the plastic. Do not shake the bowl.

13. Is the mug still empty? If not, explain why.

____

____

14. The setup you made is like a small environment. Water moves through the environment in the water cycle. Look at the diagram below.

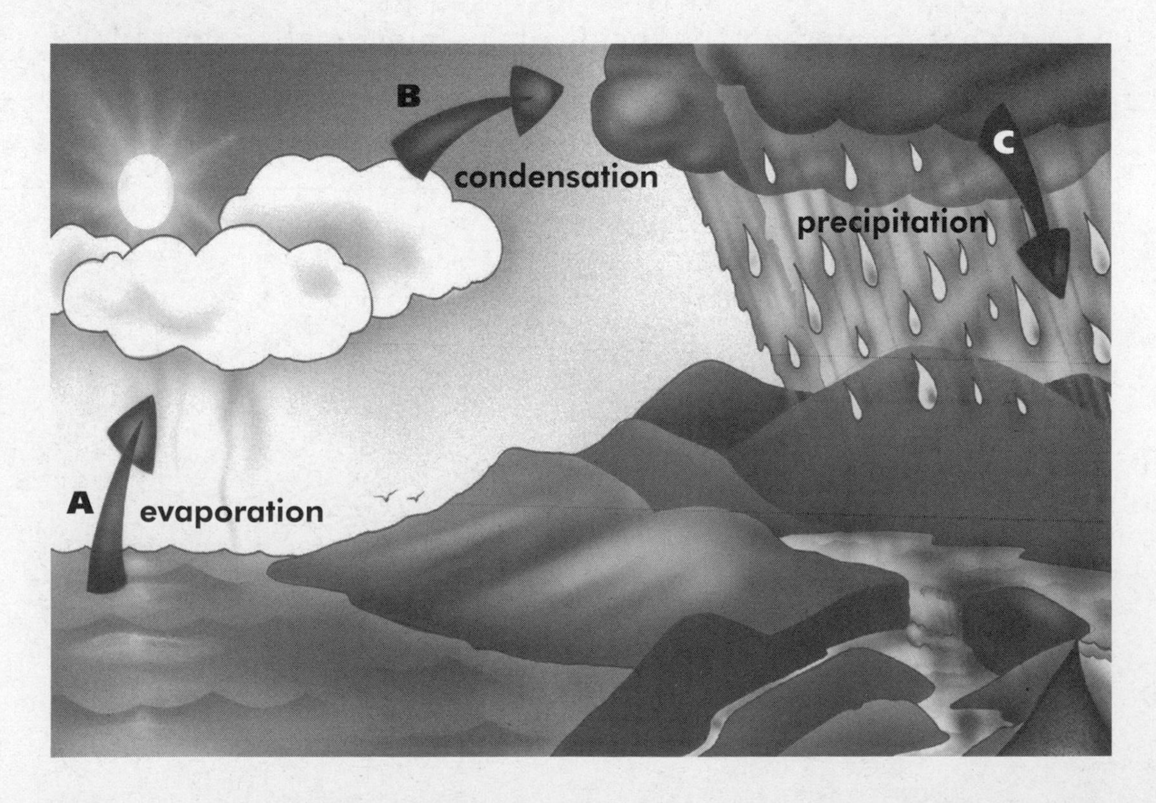

Which part of your setup is like A in the diagram? Explain.

____

____

Which part of your setup is like B in the diagram? Explain.

____

____

Which part of your setup is like C in the diagram? Explain.

____

15. When you have finished, put things back the way you found them. Wipe up any spills.

Directions (1–18): Each question is followed by four choices. Decide which choice is the *best* answer. Circle the letter of the answer you have chosen.

1 What property can help you *best* tell the difference between grains of salt and grains of sugar?

A color
B hardness
C odor
D taste

2 What happens to the volume of a gas in a closed box when the box is opened?

A The volume increases or gets bigger.
B The volume decreases or gets smaller.
C The volume stays the same.
D The gas changes to a solid.

3 A metric ruler can be used to measure the

A volume of a liquid
B circumference of a circle
C length of a stick
D curve of an oval

4 When you choose a bowling ball, which property affects its inertia?

A its mass
B the spacing of the finger holes
C its color
D the way it rolls

5 What is the classification basis for the following group: helium, water vapor, oxygen?

A All are used to fill balloons.
B All are used to light up signs.
C All are gases.
D All are unsafe to touch.

6 What property would you be testing for if you are using a circuit breaker?

A volume
B state of matter
C electrical conductivity
D magnetism

7 What is a special property of gold and silver that allows them to be made into jewelry?

A color
B hardness
C odor
D taste

8 Why do many cooking pots have wood or plastic handles?

A Wood and plastic hold up better than metal in the dishwasher.
B Wood and plastic are more comfortable than metal to hold on to.
C Wood and plastic are less heavy than metal.
D Wood and plastic do not conduct heat well.

9 Why is it useful to know special properties of some objects, such as minerals?

A It makes them more valuable.
B Knowing this makes it easier to identify the minerals.
C It makes it more fun to identify the mineral.
D all of the above

10 Nitrogen is a liquid at very cold temperatures. How can you change nitrogen to a gas?

A Warm it up.
B Make it colder.
C Decrease its volume.
D Increase its volume.

11 When you breathe out on a cold day, your hot breath hits cold air and you "can see your breath." What changes in state have occurred?

A liquid to gas
B gas to liquid
C liquid to solid
D gas to solid

12 Temperature affects how thick some liquids may be. Which liquid gets thicker when cold?

A milk
B syrup
C orange juice
D water

13 Often, in chemistry, when acid is added to water, steam is observed in the container. This is because

A the mixture splashes a bit as the acid is poured in
B the solution cools down
C heat is generated when the acid is added
D steam is always created in chemistry

14 Which has more mass, an ounce of feathers or an ounce of gold?

A feathers
B gold
C They both have the same mass.
D They both have the same volume.

15 The mass of a box of cereal can be measured with a

A
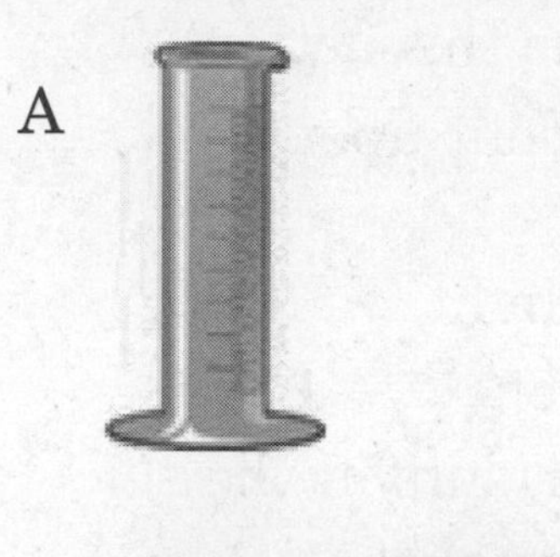

B

C
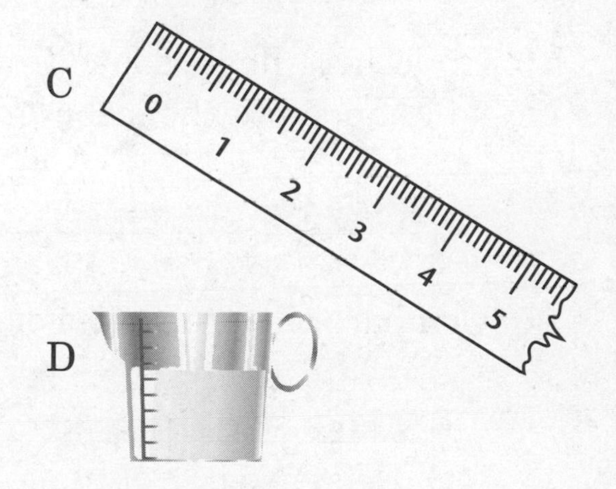

D

16 Which of the following generates heat through friction?

A car tires on the road
B erasing a pencil mark
C rubbing your hands together
D all of the above

17 The "face" of a sunflower turns throughout the day. What is this action dependent upon?

A sunlight
B water
C nutrients in the soil
D air quality

18 Which shows water changing from a gas to a liquid?

A Water on the sidewalk evaporates.
B Water droplets form on the mirror during your shower.
C Your car windows have frost on them in the morning.
D Your snowman melts.

Directions (19–25): For each question, write your answer in the spaces provided.

19 An iceberg floats south from the Arctic Ocean. What changes in state will happen to the iceberg?

____

20 Compare the properties of apples to oranges.

____

____

____

21 Magma, hot, molten rock beneath Earth's surface, often contains a lot of water. In what state of matter would you expect water to be as it erupts from a volcano?

____

____

____

22 Describe a situation where water changes from a liquid to a gas.

____

____

23 What properties would you use to describe a dog?

____

____

____

24 Classify the following sports balls based on their properties: football, soccer ball, volleyball, baseball, softball, tennis ball, golf ball.

25 Based on their properties, describe and classify the following into two groups: tomatoes, potatoes, carrots, zucchini, watermelon, cantaloupe.

Chapter 7: Energy

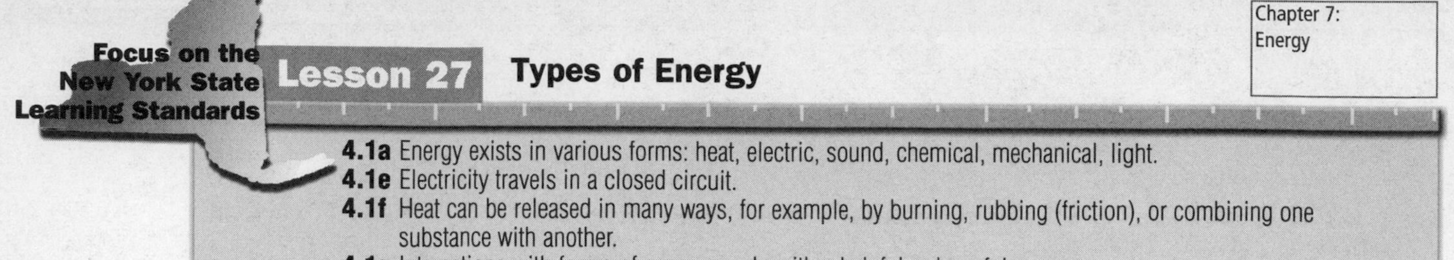

Focus on the New York State Learning Standards

Lesson 27 Types of Energy

4.1a Energy exists in various forms: heat, electric, sound, chemical, mechanical, light.
4.1e Electricity travels in a closed circuit.
4.1f Heat can be released in many ways, for example, by burning, rubbing (friction), or combining one substance with another.
4.1g Interactions with forms of energy can be either helpful or harmful.

Using different forms of energy can be helpful or harmful.

Work is done when something moves and energy is transferred.
Energy is the ability to do work.
To **vibrate** is to move back and forth rapidly.
A **circuit** is a path like a circle.

Guided Instruction

Directions Read the following information.

If you do not have **energy**, you would not be able to move. In science, energy is the ability to do **work** and work only happens when something moves. In other words, if you were reading a book, you would not be doing any work. If you were running, you would be doing work. Energy moves too. Heat energy moves from hot things to cold things. When you hold a cup of hot cocoa, the heat moves from the cup to your hands. So, when energy moves, work is being done too.

Energy exists in several forms. Heat is the energy that raises the temperature of matter. Chemical energy is the energy stored in substances such as food, gasoline, wood, or the tip of a match. Light energy moves out from objects such as the Sun or a light bulb. Sound is energy created when objects **vibrate**, causing movement in the air.

Mechanical energy is involved with moving matter. If a rolling marble strikes another, mechanical energy

Guided Questions

How are **energy** and **work** related?

What do scientists mean when they use the term **work?**

Are you working when you ride a bike?

What does an object do when it **vibrates?**

makes the second marble move.

Electric energy powers appliances such as a radio or light bulb. It travels in a closed **circuit**. Electric energy that leaves a source, such as an electric plant, must come back to its source after doing work. For example, if it comes from a source and goes to a light bulb, it must go back to that source after it lights the bulb. Otherwise, it cannot light the bulb.

Heat can be released in many ways. For example, you can burn wood, releasing heat from its chemical energy. Rubbing your hands together makes them feel warmer. Sometimes mixing one substance with another will make the mixture warm.

Releasing and using energy can be helpful or harmful. Sunlight gives energy to all living things, but it can burn your skin. Burning wood can help you keep warm or cook food, but fires can destroy forests and homes. Electric energy can make a radio work, but it can hurt or kill living things if too much of it travels through their bodies. Using gasoline can take people

Guided Questions

What is a **circuit?**

Directions For each question, write your answer in the space provided.

1. If you were pushing very hard against a brick wall, would you be doing work? Explain your answer.

____

____

2. Is work being done when water freezes into ice? Explain your answer.

____

____

3. What form of energy does an object create when it vibrates?

____

4. In what type of circuit does electricity travel when it does work?

___

5. Give an example of how energy can be harmful, and one of how it can be helpful.

Harmful: ___

___

Helpful: ___

___

6. What is one type of energy that reaches Earth from the Sun?

___

Apply the New York State Learning Standards to the State Test

Directions: Use the pictures to answer questions 7 through 11.

7 What type of energy is in the picture of the musicians?

___

8 What type of energy do you see in the picture of the windmill?

___

9 What type of energy is in the match before it is burned and while it is burning?

Before burning: ______________________________

While burning: ______________________________

10 What type of energy is traveling to the toaster?

____

11 What type of energy is traveling from the toaster?

____

Directions (12–17): Each question is followed by four answer choices. Decide which choice is the best answer. Circle the letter of the answer you have chosen.

12 If you are roasting marshmallows over a wood fire, what two types of energy are you using?

A light and heat
B heat and chemical
C chemical and light
D mechanical and sound

13 If rushing water is turning the blades of a water wheel, what form of energy is being used?

A chemical
B sound
C mechanical
D electric

14 Why is this light bulb lit?

A Electrical energy is flowing in a closed circuit.

B Light energy is flowing from the battery.

C Heat energy makes the light bulb glow.

D Chemical energy in the battery flows through the open circuit.

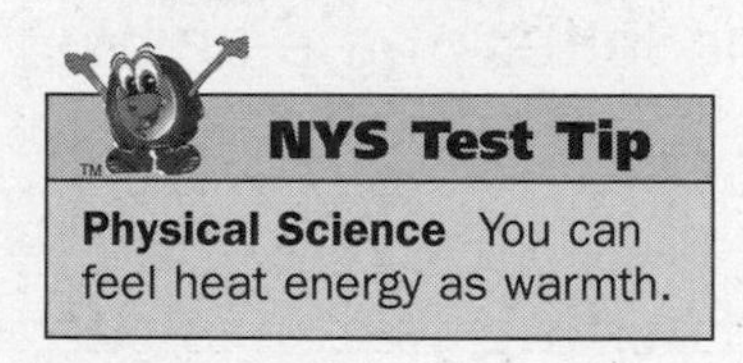

NYS Test Tip

Physical Science You can feel heat energy as warmth.

15 Electric energy can be used as

A the energy source for an electric power plant

B the light source for a solar calculator

C the chemical source for a fire

D the energy source for a refrigerator

16 Which of the following would release energy from pieces of wood?

A building a warm house with them

B rubbing them together

C leaving them in the Sun

D breaking them in smaller pieces

17 Which statement is true?

A Energy is always helpful.

B Energy can be harmful.

C Energy can be helpful and harmful.

D Energy is not helpful or harmful.

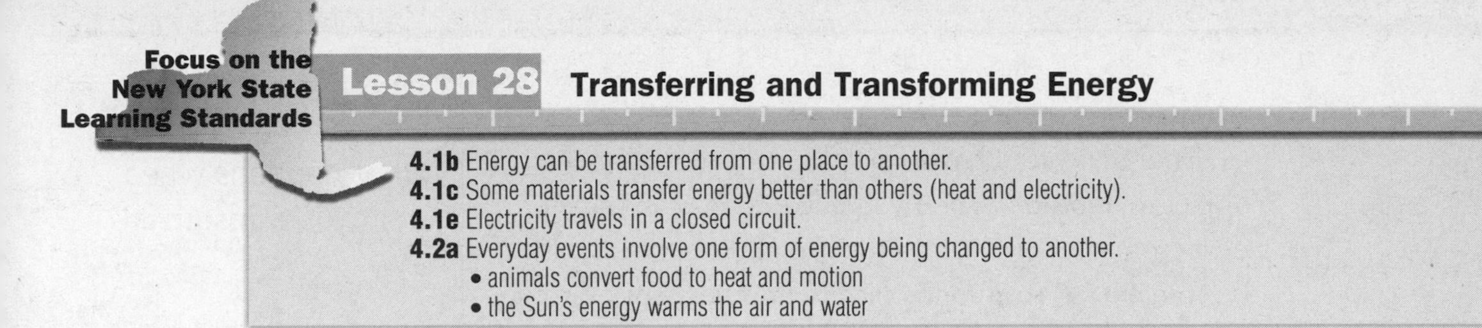
Focus on the New York State Learning Standards

Lesson 28 Transferring and Transforming Energy

4.1b Energy can be transferred from one place to another.
4.1c Some materials transfer energy better than others (heat and electricity).
4.1e Electricity travels in a closed circuit.
4.2a Everyday events involve one form of energy being changed to another.
- animals convert food to heat and motion
- the Sun's energy warms the air and water

Materials can transfer energy from one place to another.

Friction is caused by rubbing.

To **transfer** is to move from one person, thing, or place to another.

Guided Instruction

Directions Read the following information.

The first living organisms to use light energy from the Sun were plants. Plants change light energy from the Sun to food, which is stored chemical energy. When animals and people eat a plant, they change its stored chemical energy to heat to stay warm and mechanical energy to move. When plant parts such as wood or dry leaves are burned, their stored chemical energy is changed to heat energy.

When coal is burned, its stored chemical energy is changed to heat energy too. Then the heat energy may be changed to mechanical energy to turn a machine that makes electrical energy. Electrical energy can be changed into light, sound, heat, or mechanical energy. It may power a lamp, a doorbell, a hot plate, or a blender.

Mechanical energy can be changed to sound. You push a piano key and hear a musical note. Mechanical energy can also release heat. The **friction** between moving machine parts releases heat. Friction is caused by materials rubbing together. When you rub your hands together, friction makes your hands feel warm.

Guided Questions

How does energy transfer from the Sun, to plants, to you?

How does energy transfer from coal to your doorbell?

After a car is driven, why are its tires warm?

What causes **friction?**

Guided Questions

What happens when energy is **transferred?**

Energy is often **transferred** or moved from one object to another. When you make toast, you transfer heat from the toaster into the bread. Energy can also be transferred from one place to another. For example, electricity is almost always made in one place and used in another. It travels miles through wires to your home.

Some materials transfer energy better than others. Certain kinds of metals, such as copper, transfer electricity very well so they are used for electrical wire. Sometimes you do not want energy to transfer, so you might want to use material that does not transfer energy well. For example, metals transfer heat very well. That is why stoves and pots are made of metal. If a metal pot has a metal handle, the handle becomes very hot on the stove. Therefore, some pot handles are made out of wood or plastic.

Directions For each question, write your answer in the space provided.

1. What happens to electrical energy as a lightbulb lights?

____

2. What is the result of friction?

____

3. Give an example of energy transfer as you drink hot cocoa.

____

____

____

4. What energy changes takes place when coal is burned?

____

5. If you want a potato to bake faster, you can push a metal nail through its center. Why do you think the nail makes the potato bake faster?

____

____

Apply the New York State Learning Standards to the State Test

Directions: Use the pictures to answer questions 6 through 11.

1 2 3 4

6 Describe the energy transfer in picture 1.

____

____

7 What type of energy is being transferred in picture 4?

____

8 Describe the energy change in picture 4.

____

____

9 What energy change is taking place in picture 3?

____

____

10 What energy change took place when the match was struck?

____

11 Why must two wires run between a house and the electric power plant?

____

____

Directions (12–17): Each question is followed by four answer choices. Decide which choice is the best answer. Circle the letter of the answer you have chosen.

12 Before humans can use energy, they need to

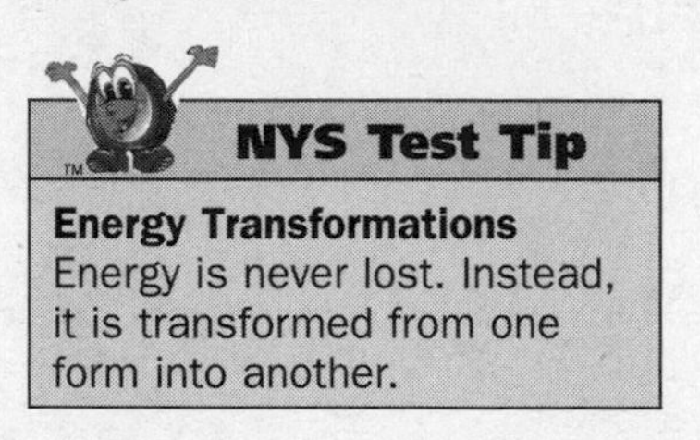

A burn something that contains stored energy
B turn it into electricity
C change it to another form
D transfer it to another place

13 The first living organisms to use light energy from the Sun were

A workers at a solar power plant
B cold-blooded animals
C warm-blooded animals
D green plants

14 What energy changes take place when the Sun shines on you?

A light to heat
B light to chemical
C light to sound
D light to mechanical

15 When a candle burns, what energy change is taking place?

A light to electrical
B electrical to chemical
C chemical to mechanical
D heat to light

16 When water is boiled in a pan on a stove, energy is transferred

A from the stove to the pan
B from the pan to the water
C from the stove to the pan to the water
D from the stove directly to the water

17 Some electrical cords have metal wires inside and rubber coating outside. The reason is that

A rubber and metal transfer electricity equally well
B rubber and metal transfer electricity better together than alone
C rubber transfers electricity better than metal
D metal transfers electricity better than rubber

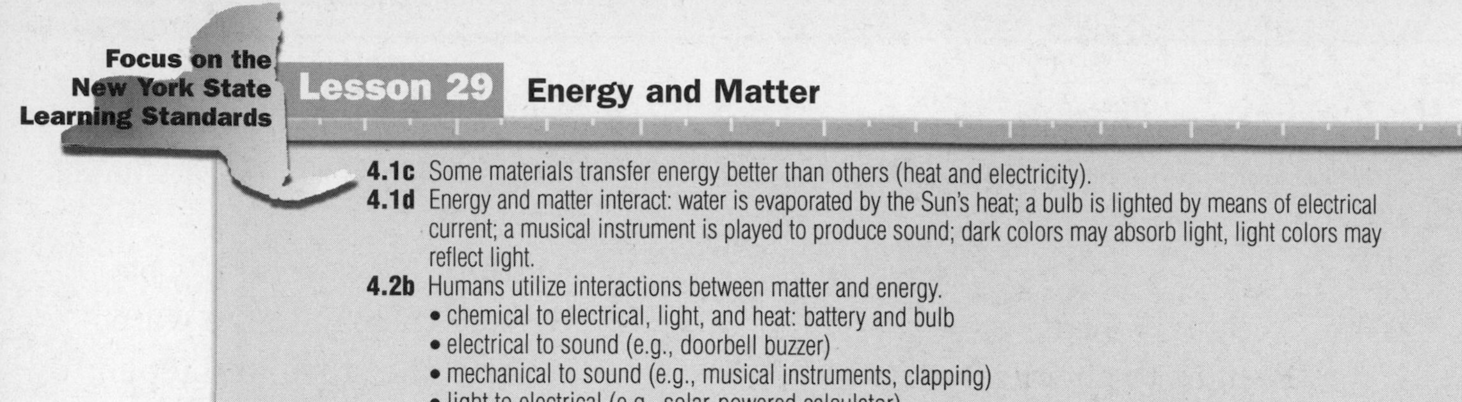

Focus on the New York State Learning Standards

Lesson 29 Energy and Matter

4.1c Some materials transfer energy better than others (heat and electricity).
4.1d Energy and matter interact: water is evaporated by the Sun's heat; a bulb is lighted by means of electrical current; a musical instrument is played to produce sound; dark colors may absorb light, light colors may reflect light.
4.2b Humans utilize interactions between matter and energy.
- chemical to electrical, light, and heat: battery and bulb
- electrical to sound (e.g., doorbell buzzer)
- mechanical to sound (e.g., musical instruments, clapping)
- light to electrical (e.g., solar-powered calculator)

Energy and matter interact and humans use interactions between matter and energy.

Batteries are containers of matter with stored energy.

Solar means powered by energy from the Sun.

Guided Instruction

Directions Read the following information.

Energy and matter interact. Energy produces changes in matter. For example, sunlight raises the temperature of water and causes it to evaporate. Matter is also used in processes that change the form of energy. When you use mechanical energy to play a musical instrument, the matter in the instrument produces sound.

Small differences in matter may cause different interactions with energy. For example, dark colors may absorb more light, while light colors may reflect more light. You may have noticed that people often wear lighter colors in the summer than in the winter. Some people paint the bottom of a swimming pool very dark. The dark bottom absorbs heat energy and transfers it to the water.

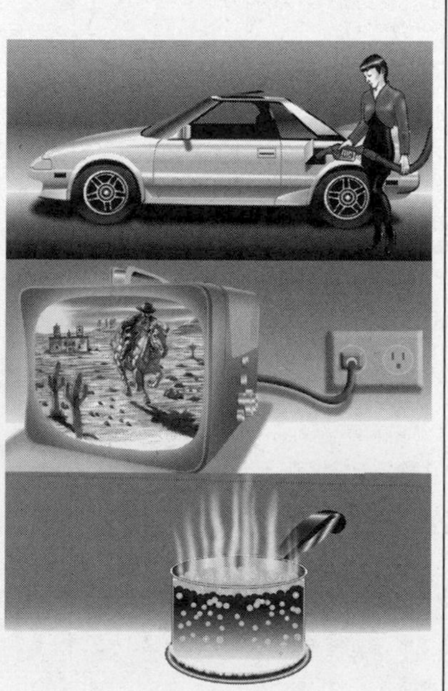

Guided Questions

You are matter, too. How does sunlight make you feel or change you?

How is energy interacting with matter, in the pictures shown?

Electrical energy may cause a bulb to light up or a toaster to heat up. In order to affect matter this way, electricity must travel in a closed circuit. The switch on a lamp opens a circuit or closes it to turn the lamp on or off. Electrical energy is sometimes stored in matter. For example, some calculators, radios, and watches run on the energy stored in **batteries**. Some **solar** batteries store energy from the Sun.

Matter is also used to transfer energy. If you want to cook fish over hot coals, you do not usually put them right on the burning coals. The burning coals heat a metal grill, and you put the fish on the grill. Suppose the electrical outlet is on one side of the room and you need a lamp on the other. You might use a long cord to transfer the electricity from the outlet to the lamp.

Some materials transfer energy better than others. Solids transfer heat better than liquids. Liquids transfer heat better than gases. For example, water transfers energy better than air. If you put an ice cube into water that is at room temperature, it will melt faster than if you leave it exposed to air at the same temperature.

Guided Questions

What are **batteries?**

What does **solar** mean?

Directions For each question, write your answer in the space provided.

1. What are solar batteries?

____

____

2. When you turn on a lamp, then turn it off, explain what happens to the circuit.

____

3. Explain how energy transferring to matter cooks an egg.

____

____

____

4. Explain why wires are covered in plastic when extension cords are made.

____

____

____

____

5. Explain whether a black or a white sun umbrella would be better to use on a hot day.

____

____

Apply the New York State Learning Standards to the State Test

Directions: For each question, write your answer in the space provided. Base your answers to questions 6 through 11 on the drawings below.

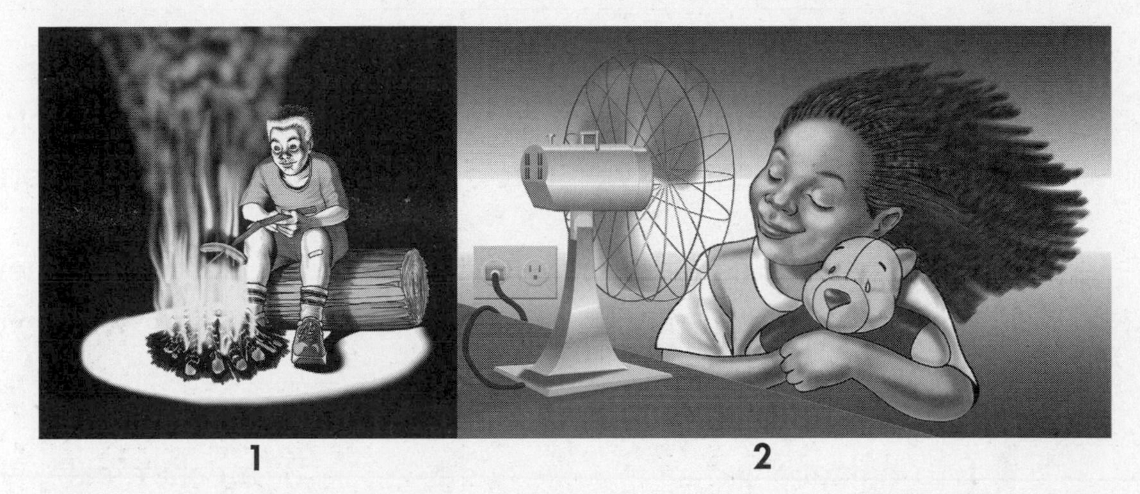

6 How is energy interacting with the matter in the wood in picture 1?

___

7 How is energy interacting with the hot dog in picture 1?

___

8 How is the boy using the interaction between matter and energy picture 1?

___

___

9 What form of energy is interacting with the cord in picture 2?

___

10 How is this same energy interacting with the fan?

___

11 How is the girl using the interaction between energy and matter picture 2?

____

____

Directions (12–17): Each question is followed by four answer choices. Decide which choice is the best answer. Circle the letter of the answer you have chosen.

12 Which contains electrical energy stored in matter?

A a light bulb
B a battery
C a toaster
D an electrical cord

13 What form of energy is in a match before you light it?

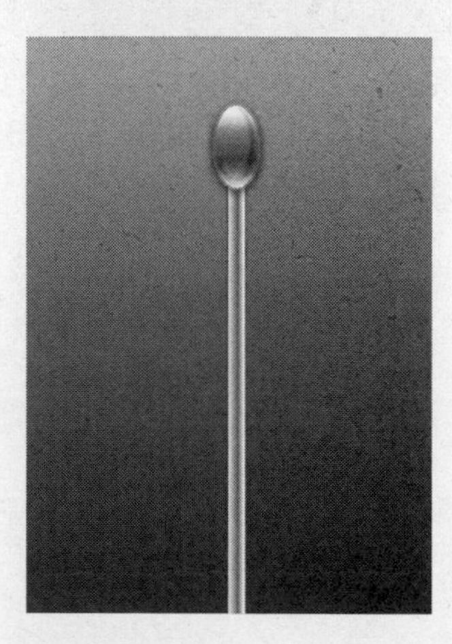

A chemical
B electrical
C heat
D light

14 What form of energy are you using when you draw?

A sound
B mechanical
C electrical
D chemical

15 Why does the plug on an electrical cord have at least two prongs?

A to be sure enough energy goes into the cord to power the appliance
B so that energy can travel from the source and back to the source
C so that neither of the holes in the outlet will be exposed
D because the electricity needs to feed an appliance from two directions

16 Jill wants a cover for her car. The weather is usually hot and sunny where she lives. What color cover would be best to keep her car from getting too hot?

A a green cover
B a red cover
C a white cover
D a black cover

17 Some materials transfer energy better than others. Which of the following statements do you think is true?

A Food will cook faster in water at 212 degrees than in an oven at 212 degrees.
B Food will cook faster in an oven at 212 degrees than in water at 212 degrees.
C Food will take exactly the same length of time to cook in water at 212 degrees as an oven at 212 degrees.
D The temperature needed to cook food in water is higher that the temperature needed to cook it in the oven.

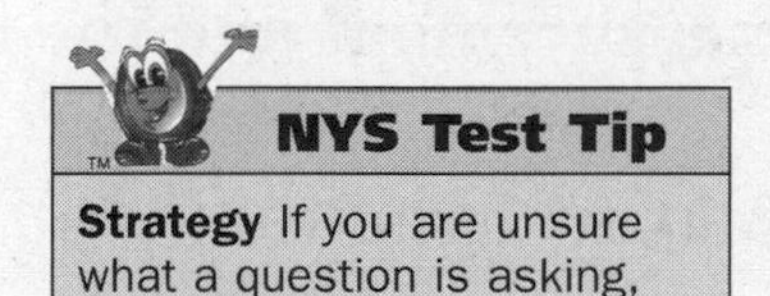

NYS Test Tip

Strategy If you are unsure what a question is asking, read it more than once.

Higher-Order Performance Task

A Dissolving Problem

Task:

You will be using different materials to see what kinds of energy affect how fast sugar dissolves. Then you will design a way to dissolve sugar in water in the least amount of time.

Materials:

- container of cold water
- container of hot water
- container of sugar
- measuring cup
- 2 clear plastic cups
- 2 plastic spoons
- 2 stirrers
- clock or watch with second hand
- container for used water
- paper towels
- 2 thermometers

Directions:

1. Use the materials listed above to complete this investigation.
2. Use a measuring cup to add the same amount of hot water to two cups. Make sure the water in each cup is the same temperature. Record the amount of water in Data Table 1 below. Give the measurement and the unit of measure.
3. Add 1 level spoonful of sugar to each cup at the same time. Do not touch or stir one of the cups. Stir the other cup. Record how long it takes the sugar to dissolve in each cup.

DATA TABLE 1

	Amount of Water	Amount of Sugar	Time to Dissolve Sugar
Hot Water, Not Stirred		1 level spoonful	
Hot Water, Stirred		1 level spoonful	

4. How does stirring affect how fast the sugar dissolves?

____

5. What kinds of energy were used to dissolve the sugar in the two cups?

____

6. Empty the cups into the container for used water. Clean up any spills.

7. Now design an investigation using the same materials to help you find out how heat affects how fast sugar dissolves. Describe your investigation. Be specific so that someone else could do your investigation. Create a data table for your results in Data Table 2 below.

____

____

____

____

DATA TABLE 2		

8. Use the materials and do your investigation. Record your results in Data Table 2.

9. What kinds of energy were used to dissolve the sugar in the two cups?

____

10. Look back at Data Table 1 and Data Table 2. How do the two kinds of energy used affect how fast sugar dissolves?

____

____

11. What would you do to dissolve sugar in water in the least amount of time?

____

____

12. Empty the cups of sugar water into the container for used water. Wipe up any spills.

Building Stamina®

Directions (1–17): Each question is followed by four choices. Decide which choice is the best answer. Circle the letter of the answer you have chosen.

1 When lightning strikes, there is an electrical discharge, and light, sound, and heat are formed. Light, sound, and heat are different forms of

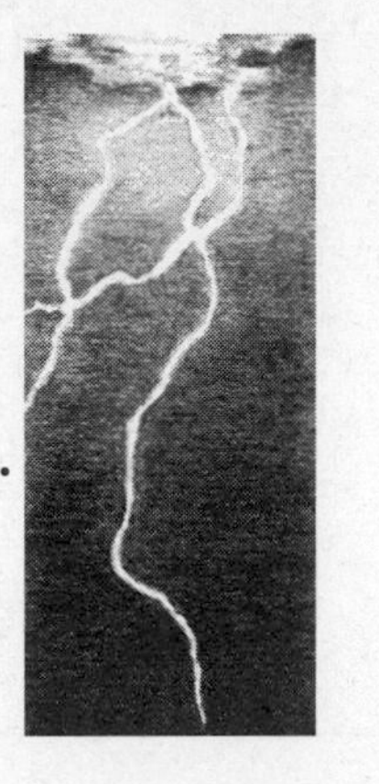

A electricity
B matter
C energy
D all of the above

2 Why does flipping a wall switch turn on a light?

A It is connected to the light bulb.
B It closes the circuit to make the electricity flow.
C It sends a signal to the bulb to light.
D It opens the circuit to stop the flow of electricity.

3 What form of energy is formed when something vibrates?

A sound
B mechanical
C friction
D chemical

4 After you eat, you go out and ride your bike. Which forms of energy did the food most likely change into?

A heat and sound
B heat and electricity
C heat and mechanical
D sound and light

5 Which can be used for heat energy without the process of burning?

A the Sun
B coal
C wood
D oil

6 Which form of energy causes your hair to stand on end when you rub a balloon on your head?

A heat
B electric
C sound
D chemical

7 You rub two sticks together and a fire starts. What is happening?

A Sound energy changes into heat energy.
B Chemical energy changes into electric energy.
C Mechanical energy changes into heat energy.
D Heat energy changes into sound energy.

8 How does the Sun's light energy change when it reaches Earth?

A It changes to heat.
B It changes to light.
C It changes to electricity.
D It changes to ozone.

9 Why do humans eat only certain plants?

A Some plants are too hard for our bodies to digest.
B We eat only the plants whose stored chemical energy can be used by our bodies.
C Some of the stored chemical energy in plants is not usable by humans.
D Some stored energy in plants needs to be changed to nutrients before we can eat them.

10 Why does burning coal produce more heat energy than burning paper?

A There is more chemical energy stored in the coal.
B Paper releases its stored energy faster than coal.
C It takes more energy to burn paper, so the effect is cancelled out.
D Coal is formed from plants, which have a lot of stored energy.

11 A strong storm caused your electricity to go out. What probably happened?

A The electric company closes down and stops the electrical output.
B Lightning got in the way of the circuit of electricity.
C The power lines got knocked down by a falling tree and the electrical circuit is no longer closed.
D Lightning energy combined with electrical energy and overloaded the system.

12 Why do potatoes cook faster when you boil them than when you bake them?

A Air doesn't transfer heat.
B Liquids transfer heat better than gases.
C Air transfers heat better than water.
D The pot transfers heat directly to the potatoes.

13 What kind of energy interacts with growing plants?

A heat
B mechanical
C light
D chemical

14 The energy in a battery allows a flashlight to turn on even though it is not plugged in. What type of energy does it use?

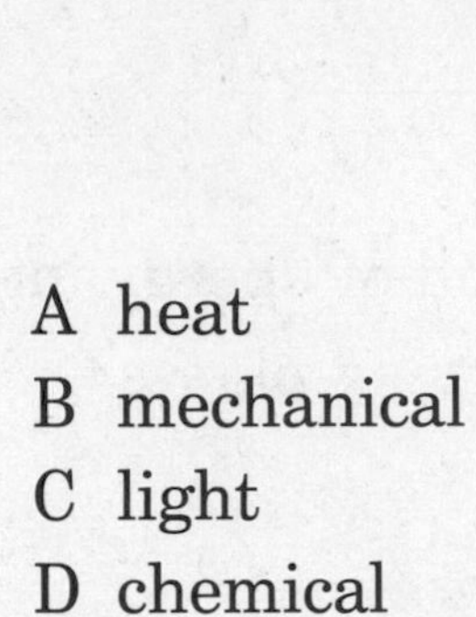

A heat
B mechanical
C light
D chemical

15 Which material would *not* make a good hotplate?

A cork
B steel
C wood
D tile

16 Why does a piece of wood get warm when you sand it with sandpaper?

A The stored chemical energy is released.
B The friction between the wood and sandpaper releases heat.
C Energy from your hand warms the wood.
D Chemical energy in sandpaper changes into heat energy.

17 Why can you put your bare hand in a 350° oven to remove a cake but you can't touch the cake pan or the cake with your bare hand?

A Our hands are used to hot air more than to hot pans.
B Only the pan and food absorb the heat in the oven.
C Our hands conduct heat energy better with metal than with air.
D The cake pan and cake transfer heat energy better than air.

Directions (18–25): For each question, write your answer in the space provided.

18. List some household kitchen items that do not use electrical energy.

19. In a piano, a small hammer hits a string to produce musical sound. In a harpsichord, a small hook plucks the string. How might the sound energy be different?

20. How does your body cause the water in a swimming pool to get warmer?

21. Why do batteries run down?

22. What are three ways in which humans use the interactions between matter and energy?

1. ____

____

2. ____

____

3. ____

____

23. What kinds of energy are present when fossil fuels are burned?

____

____

24. Loudness is measured in decibels, where the threshold of pain is 120 decibels. What does this mean?

____

____

25. How have our lives improved with the invention of gasoline? How have our lives been hurt by the invention of gasoline?

____

____

Focus on the New York State Learning Standards

Lesson 30 Mechanical Force

PS 5.1a The position of an object can be described by locating it relative to another object or the background.
PS 5.1b The position or direction of motion of an object can be changed by pushing or pulling.
PS 5.1d The amount of change in the motion of an object is affected by friction.
PS 5.1f Mechanical energy may cause change in motion through the application of force and through the use of simple machines such as pulleys, levers, and inclined planes.

Force and friction affect the motion of an object.

Force is a push or a pull on an object.

Friction is force that resists the motion of one surface past another surface.

Guided Instruction

Directions Read the following information.

Imagine you have an empty wagon that you want to move a short distance. You might push the wagon or you might pull it. Either way, you would use force to move the wagon. A **force** is a push or a pull. An object, such as the wagon, starts to move only when something pushes it or pulls on it.

If you give the wagon just a little push, it would most likely move only a little bit. If you and three friends give it a big push, the wagon would travel further. That is because the amount of force used determines how far the object moves. And if you filled the wagon with bricks, your wagon would be heavier and you would need more force to move it.

Friction also changes the motion of an object. **Friction** is the force that resists the motion of one surface past another surface. *Resist* means to fight against. Friction slows down a rolling wagon.

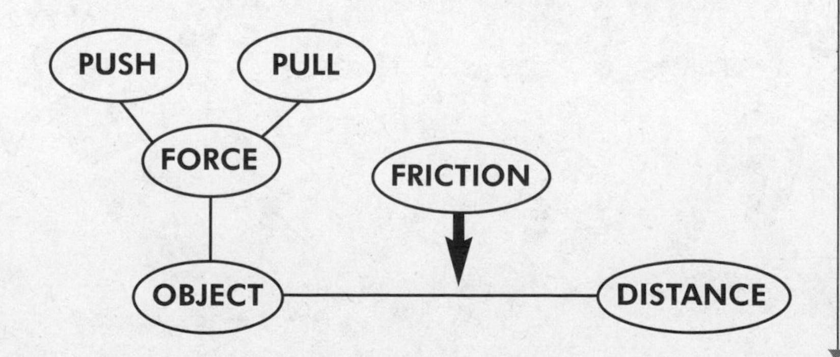

Guided Questions

What is **force?**

What is **friction?**

Explain what happens when you push a box across the floor.

Rubbing a piece of foil over wood does not have much friction. Rubbing a piece of sandpaper over wood does have friction. When you rub a piece of sandpaper back and forth across a piece of wood, the sandpaper and wood will feel warm when you touch them. When surfaces rub against each other, the roughness of the surfaces slow the movement and produce heat. The wheels on your wagon make it easier to pull it. That's because the simple machine of wheels and axles reduces friction and reduces the amount of force needed to move the wagon.

Other things, such as the mechanical energy provided by simple machines, also affect the motion of an object. Pushing the wagon uphill would take more force than pushing the wagon downhill. The hill acts like a simple machine called an inclined plane, which changes the motion of the object.

Guided Questions

Directions For each question, write your answer in the space provided.

1. What do you need to do to make a wagon move?

____

2. Look at the drawing below. What happens in a tug of war when both sides are pulling with equal force?

____

3. If another student joins the game and stands behind the two students already in the game, how will the game be changed?

____

____

4. Which would take more force: moving an automobile one mile or moving it ten miles? Why?

____

____

____

5. Is it more difficult to ride a bicycle uphill or on a level surface? Why?

____

____

____

6. Describe and compare what forces are needed to make a bicycle move and stop.

____

____

____

____

Apply the New York State Learning Standards to the State Test

Directions (7–10): For each question, write your answer in the spaces provided. Base your answers to questions 7 through 10 on the paragraph below.

Imagine pushing an empty sled over ice. The sled is light in weight and you can move it easily by yourself. Also, the smooth, slippery ice makes the work of pushing the sled quite easy, because there is little friction between the smooth sled runners and the surface of the ice.

7 Suppose you are pushing the same sled over a rocky road. Why will it take more force to push the sled over a rocky road than over ice?

____

____

____

8 Two friends come along and want to ride on the sled. Explain whether it will take more or less force to push the sled over ice with two friends riding on it.

____

____

9 You decide to go sledding on a snow-covered hill. Which will take more force: to pull the sled over level ground to the bottom of the hill or to slide downhill from the top of the hill to the bottom? Why?

____

____

____

10 Which would take more force to move over ice: a sled with a rough metal bottom or a sled with a smooth metal bottom? Why?

____

____

____

Directions (11–16): Each question is followed by four choices. Decide which choice is the best answer. Circle the number of the answer you have chosen.

11 What will happen if you push or pull a small object?

A The object will move.
B The object will not move.
C The object will float.
D The object will disappear.

12 On which path would it take the least amount of force to move a bicycle forward?

A uphill path
B downhill path
C straight and level path
D upward spiral path

13 The least amount of force would be used to pull a sled along a path that is covered with

A grass
B sand
C ice
D stones

14 Which of the following best reduces the force needed to move a wagon?

A wheels
B a handle
C metal sides
D painting it red

15 When sandpaper is rubbed over wood, both surfaces become warm. This is a result of

A tension
B fraction
C suction
D friction

16 Which word completes this diagram to explain the movement of an object?

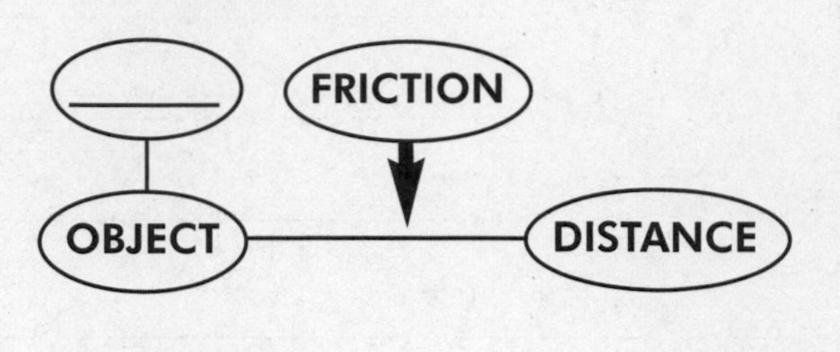

A friction
B wagon
C force
D motion

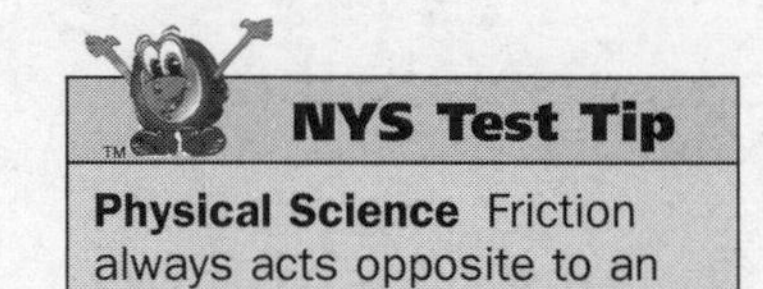

Physical Science Friction always acts opposite to an object's motion.

Focus on the New York State Learning Standards

Simple Machines

PS 5.1f Mechanical energy may cause change in motion through the application of force and through the use of simple machines such as pulleys, levers, and inclined planes.

Six simple tools make work easier.

Simple machines make work easier by changing the strength, direction, or speed of a force.

Mechanical energy is the energy an object has because of its motion and the forces acting on it.

Directions Read the following information.

Tools with only one or two parts are known as **simple machines**. Simple machines use **mechanical energy** to change the strength, direction, or speed of a force. Work, such as lifting, cutting, prying, tightening, and moving objects, is easier when you use simple machines. Here are six simple machines:

Inclined Plane

A smooth board is a plane. When the board, or plane, is slanted, it can help you move objects across distances. A ramp is a common inclined plane. Moving a heavy box is easier if you slide the box up or down a ramp.

Wedge

When you use the pointed edges of an inclined plane to push things apart, the inclined plane is a wedge. A chisel, when used to split a piece of wood, is a wedge.

Screw

An inclined plane wrapped around a cylinder becomes a screw. Every turn of a metal screw helps you move a piece of metal through a wooden space.

Guided Questions

What are **simple machines?**

Lever

A tool that pries something loose or that lifts with an arm-like motion is a lever. A shovel or a playground seesaw can be a lever.

Wheel and Axle

Another kind of simple machine is the wheel and axle. The wheel turns the axle, which causes movement. For example, on a wagon, the metal wagon bed rests on top of the axles. The wheels below rotate on the axle and the wagon moves.

Pulley

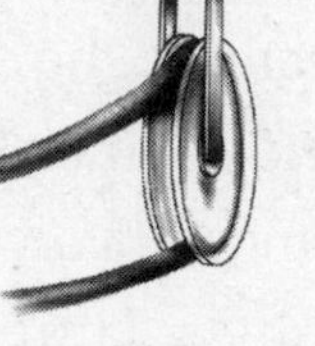

The wheel can also rotate a rope. This is a pulley. In a pulley, a rope wraps around a wheel. As the wheel rotates, the rope will move. The rope can be used to raise and lower objects. For example, a flag on a flagpole is raised and lowered by a pulley.

Guided Questions

Where is the wheel and axle on a wheelbarrow?

What are the names of the six simple machines?

Directions For each question, write your answer in the space provided.

1. Which two simple machines would be useful if you wanted to move a heavy box up a stairway?

____

2. Which simple machine would be useful to attach two pieces of wood together?

____

3. Which simple machine would you find on a wagon, a car, and a truck?

____

4. Which three simple machines are based on inclined planes?

____

5. Do you think a baseball bat and a tennis racket are simple machines? Why or why not?

____

____

6. If an axe head is a wedge, are scissors a pair of wedges? Why or why not?

____

____

Apply the New York State Learning Standards to the State Test

Directions: For each question, write your answer in the spaces provided. Base your answers to questions 7 through 11 on the figures below.

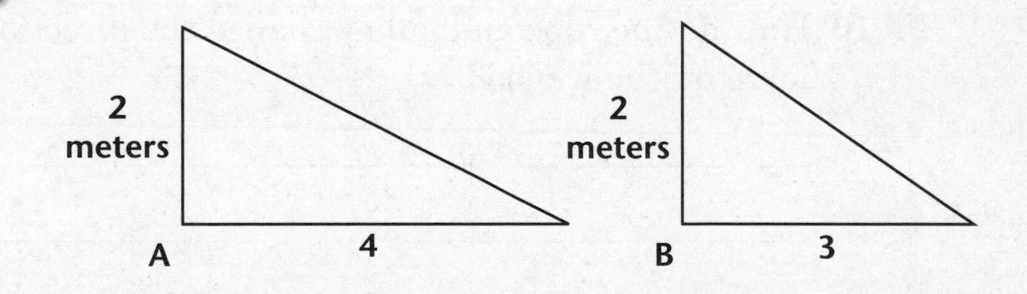

7 Two ramps lead to a door at the top of a high step. Both ramps are 2 meters high, but Ramp A is 4 meters long and Ramp B is 3 meters long. You need to push or pull a wagon to the door. Which ramp would make your work easier? Explain your answer.

____

____

8 How might you use a rope and pulley to move the wagon up the ramp?

____

____

____

____

9 How could you use a rope and pulley without the ramp to get the wagon to the door? Where would the pulley need to be attached? Explain or draw your answer.

____

____

____

10 How do the rope and pulley change the direction of the force on the wagon?

____

____

11 Explain or draw how you could use a lever to lift a rock.

____

____

____

Directions (12–17): Each question is followed by four answer choices. Decide which choice is the best answer. Circle the letter of the answer you have chosen.

12 What are tools with few or no moving parts called?

A force
B work
C simple machines
D compound machines

13 What are simple machines?

A machines that make work harder
B machines that make work easier
C machines that make work more expensive to complete
D machines that make work impossible to do

14 Which simple machine helps you move your bike up stairs?

A screw
B wheel and axle
C wedge
D inclined plane

Use the diagram to answer questions 15 and 16.

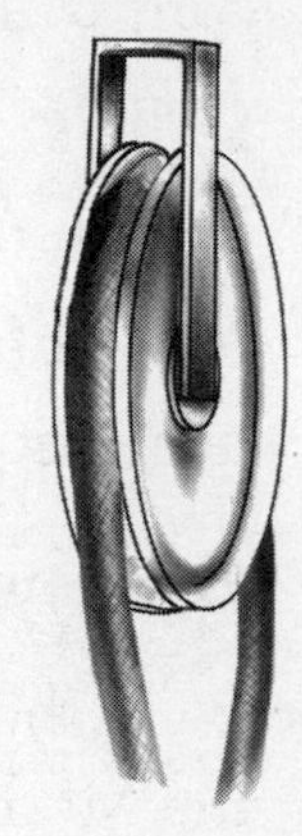

15 What simple machine is shown in the diagram?

A lever
B screw
C pulley
D inclined plane

16 What action would you take to raise a box attached to one end of the rope?

A Pull down on the free end of the rope.
B Push up on the free end of the rope.
C Swing the box back and forth.
D Twist the box.

Focus on the New York State Learning Standards

Lesson 32 Gravity and Magnetism

PS 5.1c The force of gravity pulls objects toward the center of Earth.
PS 5.1e Magnetism is a force that may attract or repel certain materials.
PS 5.2a The forces of gravity and magnetism can affect objects through gases, liquids, and solids.
PS 5.2b The force of magnetism on objects decreases as distance increases.

You can describe the effects of gravity and magnetism.

Gravity is a force of attraction between two objects.

Magnetism is a force that may attract or repel certain materials.

Magnetic field is the space around a magnet where the force of attraction is felt.

Guided Instruction

Directions Read the following information.

In the 17th century, Isaac Newton wondered why the Moon orbits Earth. He also wondered why apples fall from apple trees. What Newton discovered was the force called **gravity**. Gravity is a force of attraction between objects. It pulls apples toward the center of Earth and it also keeps the Moon in orbit around Earth.

The Moon doesn't fall to Earth like an apple. If there were no gravity, the motion of the Moon would be a straight path away from Earth. But the pull of gravity causes the path of the Moon to curve around Earth. The Moon has gravity too. Because the Moon is smaller than Earth, its gravity is less than Earth's.

Gravity works through gases, liquids, and solids. Air stays around Earth because of gravity. Oceans do not fly off into space because of gravity. Rocks and soil stay on Earth because of gravity. You stay on Earth because of gravity too. Without gravity, gases, liquids, and solids would not be pulled to the center of Earth.

Guided Questions

What is **gravity?**

What keeps the Moon in a path around Earth?

Magnetism is the force that may attract certain materials. Iron or materials with iron in them such as paper clips and iron nails are attracted to a magnet. And, like gravity, the force of magnetism can have this effect through gases, liquids, and solids. You know a magnet can hold paper on a refrigerator. The magnetic force is going through the solid paper. If you use the magnet to hold two pieces of paper, the magnet will still hold the paper on the refrigerator. What happens if you try to use the magnet to hold ten pieces of paper? The magnet probably would not work. That is because the force of magnetism on objects decreases as the distance increases.

The **magnetic field** of a magnet is the space around the magnet where its force, or magnetism, can be felt. If you lay a piece of clear plastic over a magnet and sprinkle iron filings on the plastic sheet, the filings line up in a pattern of curved lines as shown in the diagram below. The filings make the pattern because the magnetic field is strongest near the ends, or poles, of the magnet.

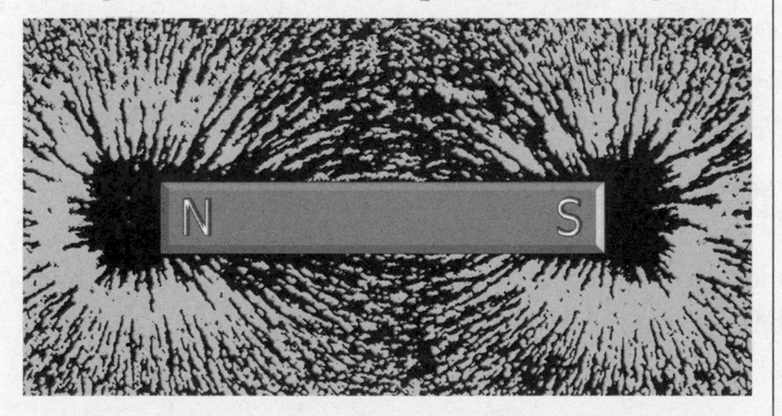

If a magnet is hung so that it can move freely, one pole will point north. That is because Earth itself is a large magnet. Magnets have two poles, a north pole and a south pole. If you placed two magnets side by side, you would see that opposite poles attract, or come together, and like poles repel, or move apart from, each other.

Guided Questions

What is **magnetism?**

What happens when a nail gets close to a **magnetic field?**

Where is the magnetic field the strongest?

Directions For each question, write your answer in the space provided.

1. How are falling apples and the orbit of the Moon alike? How are they different?

   Alike: ______________________________

   Different: ______________________________

2. How would the motion of the Moon be different if there were no gravity?

   ____

   ____

3. Give an example that shows that gravity affects objects through gases, liquids, and solids.

   Gases: ______________________________

   Liquids: ______________________________

   Solids: ______________________________

4. How are the forces of gravity and magnetism alike?

   ____

5. If you sprinkle iron filings on a plastic sheet that is placed over a magnet, what will you see? Draw your answer.

6. If you use a magnet to pick up steel pins, where on the magnet would you expect the most pins to stick? Why?

____

____

Apply the New York State Learning Standards to the State Test

Directions: For each question, write your answer in the spaces provided. Base your answers to questions 7 through 10 on the paragraph and table below.

Your class asks the following question: Are all kinds of objects attracted to a magnet? To test this question, you gather several items and a bar magnet and test each object for its attraction to the magnet. Then you record the data in the following table:

OBJECT	ATTRACTED TO THE MAGNET?
staple	yes
rubber eraser	no
string	no
safety pin	yes
toothpick	no
aluminum foil	no
copper penny (one cent)	no
silver ring	no
paper clip	yes

7 What conclusion can you make about the objects in the table?

____

8 What conclusions about metal objects can you make from the information in the table?

____

9 Explain why some metal objects are not attracted to the magnet.

____

10 Magnets attract some common objects because they are made of steel, which is mostly iron. From the information in the table, which objects might be made of steel?

____

Directions (11–16): Each question is followed by four choices. Decide which choice is the best answer. Circle the letter of the answer you have chosen.

11 What is gravity?

A the space around a magnet where the force of attraction is felt
B a force that may attract or repel certain materials
C a force of attraction between two objects
D Isaac Newton's middle name

12 What keeps the Moon in orbit around Earth?

A magnetism
B gravity
C a magnetic field
D radiation from the Sun

13 Through which of the following can magnetic force pass?

A gases
B solids
C liquids
D all of the above

14 Which of the following objects would be attracted to a magnet?

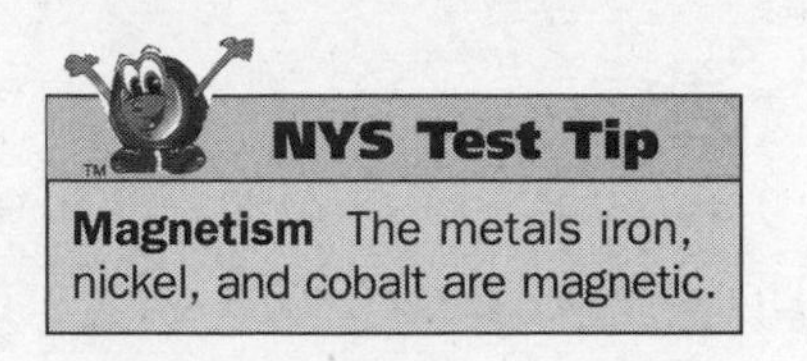
NYS Test Tip
Magnetism The metals iron, nickel, and cobalt are magnetic.

A a shoelace
B a refrigerator door
C a roll of 100 pennies
D a silver necklace

15 Why does a baseball hit up into the air lose force and fall back to Earth?

A The baseball was not hit hard enough.
B The force of gravity causes the baseball to fall to the ground.
C The magnetic outfield attracts the baseball downward.
D It was a foul ball.

The chart below lists some objects and shows if they are picked up by a bar magnet.

OBJECT	ATTRACTED TO THE MAGNET?
eraser	no
string	no
steel wire	yes
pin	yes
staple	yes

16 Which conclusion can be drawn from the information in the chart?

A The eraser was picked up by the magnet.
B The magnet picked up every object.
C The metal objects can be picked up by a magnet.
D The rubber objects can be picked up by a magnet.

Higher-Order Performance Task

Traveling the Farthest

Task:

You will find out how far a toy car moves when it travels down a ramp. Then you will design an experiment to find out how you can make the toy car go the farthest distance.

Materials:

- toy car
- rectangular board for a ramp
- 3 thick books
- weights (washers)
- masking tape
- meterstick

Directions:

1. Use the materials above to complete this task.

2. Make a ramp by placing one end of the board on one of the books. Place your ramp on the floor so that the toy car will not run into anyone or anything after it leaves the ramp.

3. Place the car at the top of the ramp and release it. Put a piece of masking tape on the floor where the car stops.

4. Use the meterstick to measure how far the car traveled from the top of the ramp.

5. Record this measurement in *Data Table 1* below. Be sure to record what units you used to measure the distance.

DATA TABLE 1	
NUMBER OF BOOKS	DISTANCE CAR TRAVELED
1	

6. Study the materials and think about how you could use them to make the toy car go the farthest distance.

7. Design an experiment using these materials. You can experiment with more than one variable, but make sure you are only testing one variable at a time. For example, you can raise the ramp, then test the car. You only changed one variable—the height of the ramp. If you raised the ramp and changed the kind of car you used, you are changing two variables and then your conclusions will not be accurate.

8. What variable or variables are you going to test and how are you going to test them?

____

____

____

9. A control is the first way you experimented. All the variables after that are compared to the control. What is the control in your investigation?

____

____

10. Create a data table for recording your results in the space below. Label it *Data Table 2*.

NUMBER OF BOOKS	NUMBER OF WEIGHTS	DISTANCE CAR TRAVELED

11. Do your experiment and record your results in your data table.

12. Explain how you made the toy car go the farthest distance.

____

____

____

____

13. Put the materials back the way you found them. Remove any tape on the floor and dispose of it properly.

Building Stamina®

Directions (1–18): Each question is followed by four choices. Decide which choice is the best answer. Circle the letter of the answer you have chosen.

1 Where does the force of friction act on a bicycle?

A tires and brakes
B tires and pedals
C seat and tires
D brakes and seat

2 Why do astronauts train underwater?

A Working in water slows you down, just as gravity in space does.
B Water makes you feel like you are wearing bulky equipment like astronauts wear.
C You are buoyant in water, giving a closer feeling to weightlessness in space.
D Holding your breath under water simulates low oxygen just like in space.

3 What simple machine is used to stop a bicycle with a hand brake?

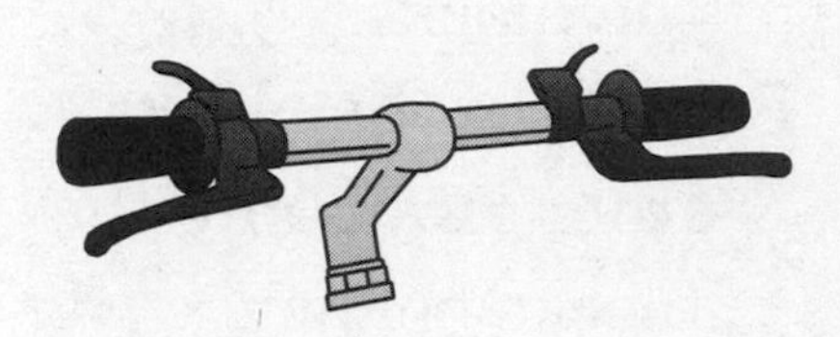

A inclined plane
B lever
C screw
D pulley

4 The Greek inventor Archimedes used a coil-shaped device that turned and brought water up from wells. This simple machine was a

A screw
B wedge
C lever
D pulley

5 Why might the builders of the Great Pyramid in Egypt have used oil beneath the large blocks that make up the pyramid?

A to reduce friction beneath the blocks
B to keep the blocks from scratching the blocks below
C to help keep the builders from getting rope burns
D to add color to the blocks

6 Blocks that make up the Great Pyramid were raised to great heights. Which simple machine may have helped the most?

A inclined plane
B wedge
C lever
D pulley

7 You are trying to attract a magnetized ball with one end of a bar magnet. The ball will not stick. Why?

A The ball is magnetized north and you are using the south end of the bar magnet.
B The ball and end of the bar magnet are both magnetized north.
C The strength of the ball is too great and repels the bar magnet.
D Both ends of the bar magnet are magnetized north so the repelling is even stronger.

8 Water expands when frozen. Rocks often split apart as ice forms in their cracks. What kind of simple machine is similar to this action?

A inclined plane
B wedge
C lever
D pulley

9 Some kites have more than one string. What do these strings do?

A They help the kite fly higher.
B Extra string helps reduce the air friction.
C Extra string helps the kite get into the air.
D Pulling on a string steers the kite in a different direction.

10 Years ago, homes without indoor plumbing used a hand pump to bring water up from their wells. The pump handle was a

A screw
B wedge
C lever
D pulley

11 A laundry line can be hung between two posts using a simple machine to form a continuous circle. As you pull on one string, the line moves to bring the clothes closer to you. Which simple machine should be used?

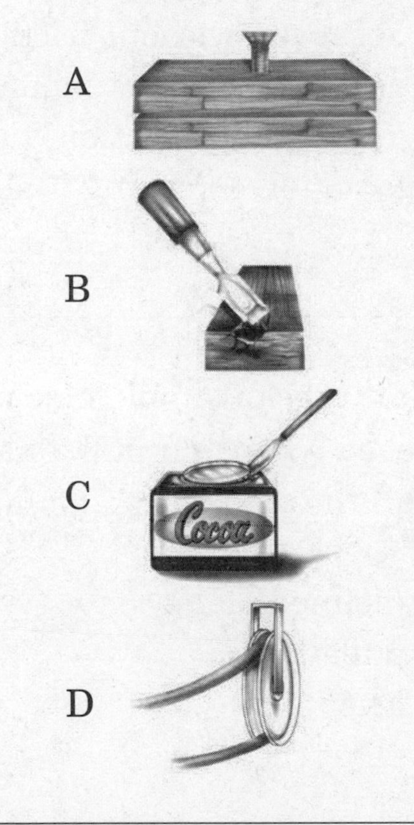

12 You are running a race with four other people. You are about to overtake the second-to-last runner. What position are you in?

A last place
B fourth place
C first place
D third place

13 If there were no air friction, would a thrown ball still fall to the ground?

A No, it would keep going at the same speed.
B No, without friction, it can actually speed up.
C Yes, even without friction, gravity is still acting on it.
D Yes, without friction, there is no force keeping it aloft.

14 What type of simple machine is used when you pull a nail from the wall with a hammer?

A inclined plane
B pulley
C lever
D all of the above

15 Why does a thrown ball slow down before it eventually falls to the ground?

A No one has enough strength the keep a ball going forever.
B The ball is affected by the force of Earth's rotation.
C The ball creates friction with the air.
D The gravitational pull of the Sun causes the ball to slow down.

16 Look at the drawing. What causes both Earth and the Moon to circle the Sun?

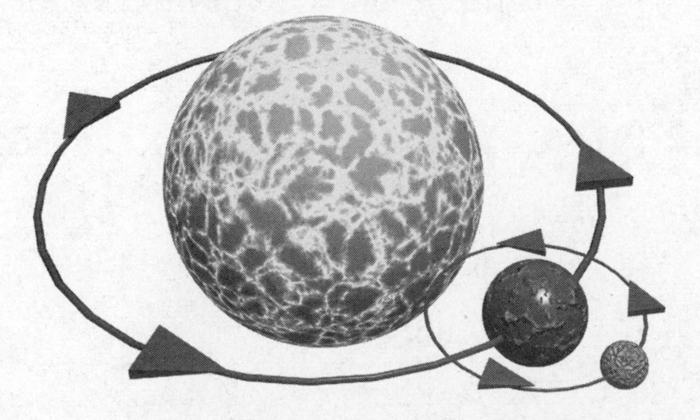

A the Moon's gravity
B the Earth's gravity
C the Sun's gravity
D Mercury and Mars force Earth and the Moon to circle the Sun.

17 You use a spoon to pry off a jar lid. What simple machine did you use?

A wheel and axle
B inclined plane
C lever
D pulley

18 You decide to ride on the swings at a park. What makes the swings go up and down?

A Push or pull makes it go up and gravity makes it come down.
B Gravity makes it go up and down.
C A pulley makes it go up and gravity makes it come down.
D A push makes it go up and friction brings it down.

Directions (19–23): For each question, write your answer in the spaces provided.

19 Earth has a strong magnetic field and is essentially a big magnet. Where are the magnetic fields strongest on Earth?

____

____

20 Where there is no atmosphere, there is no air friction. The Moon has no atmosphere. What kind of motion would you expect if you threw a ball on the Moon?

____

____

21 The force of gravity is less on the Moon than it is on Earth. Explain how your ability to jump could change on the Moon.

____

____

22 How do scientists know that Earth's gravity and magnetism can work through gases, solids, and liquids?

____

____

23 If a wheel and axle help objects move better, why are there no animals with wheels?

____

____

____

End-of-Book

Building Stamina®

The end-of-book **Building Stamina®** is a comprehensive review of all the New York State Learning Standards and Major Understandings for Science covered in the lessons. By practicing with these challenging, broad-based, higher-level thinking questions, you will be building up your stamina to succeed on the New York State Grade 4 Science Test and in other academic endeavors that require higher-level thinking.

Directions (1–95): Each question is followed by four choices. Decide which choice is the best answer. Circle the letter of the answer you have chosen.

1 Which of the following is a producer in the food chain?

A deer
B lion
C bean plant
D farmer

2 A scientist would need which sense to make an observation?

A sight
B smell
C hearing
D all of the above

3 A student placed one hundred dominoes upright in a row. He knocked over the first domino and the rest all fell over, one after another. What made the domino at the end of the row move?

A chemical energy
B heat energy
C mechanical energy
D electric energy

4 What is something in this picture that does *not* live and thrive?

A flower
B rabbit
C girl
D camera

5 A student poured milk into a glass filled with water by mistake. Liquid spilled over the sides of the glass. What conclusion can be made from this observation?

A Milk and water do not mix.

B Two objects cannot be in the same space at the same time.

C Two objects can be in the same space at the same time.

D Sometimes two objects can be in the same space at the same time.

6 What is the correct order of the stages in the life cycle of a bean plant?

A seed → seedling → mature plant

B flowering tree → seed → mature plant

C seed → seedling → apple

D none of the above

7 An example of a predator and prey is a

A deer and grass
B wolf and grass
C wolf and deer
D rabbit and carrot

8 Gina is wearing shorts and a tee shirt. She is struggling to hold onto an umbrella that is inside out. The weather is

A cool, rainy, cloudy, and calm

B cool, rainy, cloudy, and windy

C warm, rainy, cloudy, and windy

D warm, rainy, cloudy, and calm

9 A strong taste and smell are properties of an

A onion
B apple
C egg
D apricot

10 Which is an example of a food chain?

A wolf → grass → deer → worm

B tomato → deer → grass → wolf

C grass → deer → wolf

D none of the above

11 What change in state happens when wet clothes are dried in a hot clothes dryer?

A A solid changes to a liquid.

B A liquid changes to a solid.

C A liquid changes to a gas.

D A gas changes to a liquid.

12 A student reads that a basil plant can sometimes have a sweet odor. How can she make a similar observation?

A by smelling a basil plant

B by touching a basil plant

C by boiling basil leaves

D by dipping basil leaves into a bowl of sugar

13 The chart below shows the air temperature at 9 A.M. for three days of one week. What was the average temperature for those three days?

DAY	TEMPERATURE AT 9 A.M.
Sunday	15°C
Monday	19°C
Tuesday	17°C

A 17°C

B 15°C

C 16°C

D 18°C

14 What is the niche of a decomposer in the food chain?

A eats fresh grass

B uses dead organisms for nutrients

C hunts and eats deer

D receives energy from the Sun

15 The symbols on the New York map below show the weather in various parts of the state. Based on the weather map, New York's weather is

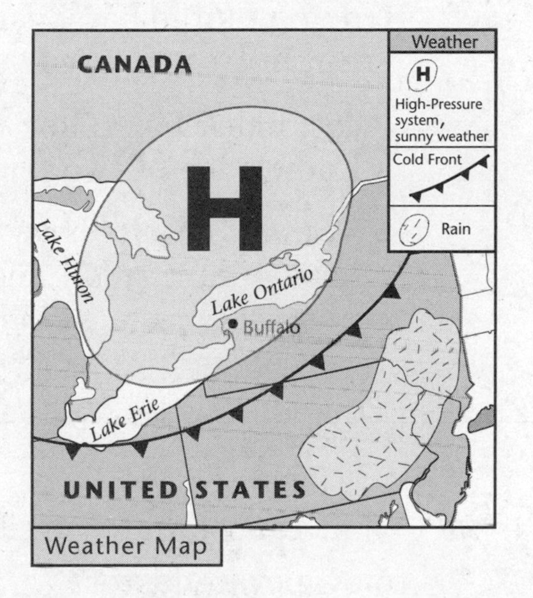

Weather Map

A rainy in the west and sunny in the east

B sunny in the west and rainy in the east

C stormy throughout the state

D sunny everywhere

16 Which of the following observations can be tested?

A Plants grow toward the Sun.

B Plants need water to grow.

C Paperclips are attracted to magnets.

D all of the above

17 Which of these is not a life process for a bird?

A hibernating
B breathing
C laying eggs
D eating

18 The most water evaporates from

A ponds
B lakes
C oceans
D rivers

19 When a person takes a bite out of an apple, the apple's mass

A stays the same
B becomes twice as large
C becomes less
D becomes greater

20 How does a green plant get food?

A A green plant makes its own food using the Sun's energy.
B A green plant soaks up food from the ground.
C Dead animals feed the plant.
D Flowers on a plant make food for the plant.

21 Which property of a rubber band changes when it is stretched?

A color
B mass
C size
D state

22 Choose the list that contains a species and its inherited traits.

A orange: orange color, round shape, bruised
B dog: pointed ears, yellow fur, brown eyes, curly tail
C cat: black hair, short hair, knowledge of two tricks
D human: brown eyes, brown hair, small nose, pierced ears

23 What causes water on Earth to evaporate faster?

A clouds
B the Sun
C rain
D sleet

24 Which of the following could be from an experiment to investigate the movement of a plant in relationship to sunlight?

A Materials: cereal, soup, juice;
First Step: read nutrition label;
Record: energy level, food pyramid group

B Materials: soil, seeds, clay pot;
First Step: plant seeds into clay pot filled with soil;
Record: growth, water added

C Materials: mature sunflower plant;
First Step: place plant near an outside window;
Record: time of day, weather, movement

D Materials: ruler, sunflower;
First Step: measure height of sunflower;
Record: height and date

25 What are the first two things a seed begins to grow after it has germinated?

A branches and flowers

B roots and a stem

C leaves and new seeds

D roots and new seeds

26 Which shows stages of the water cycle in the correct order?

A evaporation → runoff → groundwater

B condensation → evaporation → groundwater

C condensation → runoff → groundwater

D evaporation → condensation → precipitation

27 A winding road up a mountain is actually an inclined plane. Why is it easier for a car to travel on a road that winds around a mountain than on a road that goes straight up one side of the mountain?

A The car uses more force.

B The car uses less gas.

C The car uses less force.

D The car uses less water.

28 A trait that is *not* inherited by offspring is

A broken branches on an elm tree

B webbed feet on a duck

C red wings on a cardinal

D yellow petals on a tulip

29 Which of the following is transferred from plants to animals through the food chain?

A air

B the Sun's energy

C decomposers

D water vapor

30 Which action produces the most heat during a baseball game?

A catching a baseball

B walking to home plate

C throwing a baseball

D sliding into first base

31 A turtle's shell provides help for which of the following functions?

A growth

B survival

C reproduction

D walking

32 The wood cube and the sponge cube shown on the balance below are exactly the same size. What can you conclude by looking at the balance?

A The wood cube and sponge cube have equal masses.

B The sponge cube has more mass than the wood cube.

C Objects of equal size always have the same mass.

D Objects of equal size do not necessarily have the same mass.

33 The plants in a farmer's field are drying up. What part of the water cycle needs to happen to save the plants from dying?

A precipitation
B evaporation
C runoff
D condensation

34 What are the three main things a plant needs to grow?

A sunlight, water, air
B fertilizer, water, shade
C sunlight, water, bread
D water, air, heat

35 A magnet held near iron filings attracts many of them. A magnet held farther away from iron filings attracts fewer of them. This happens because

A magnetism decreases as distance decreases
B magnetism increases as distance increases
C magnetism decreases as distance increases
D magnetism causes gravity to decrease

36 What might happen to an area where lots of trees started to grow in a field of grass?

A Grass and ferns would grow stronger and deer would thrive.
B Grass would die, deer would leave, and squirrels would move in.
C Grass would grow, and wolves would move in and eat the deer.
D The trees would die, ferns would overtake the grass, and squirrels would move into the area.

37 Growing, eating, breathing, reproducing, and eliminating waste are basic life functions of a bird. Which feature of a bird would help with both eating and breathing?

A bill
B feathers
C claws
D eyes

38 What is the life cycle of a moth?

A egg → larva → pupa → adult moth
B egg → tadpole → adult frog → adult moth
C egg → pupa → larva → metamorphosis
D egg → hatchling → young turtle → adult turtle

39 Which of these statements best describes a tomato plant?

A It hunts for its food.
B It is supported by thick bark.
C Its green leaves gather sunlight.
D all of the above

40 Which interaction with mechanical energy can be harmful to humans?

A wind moving the sails of a sailboat
B wind moving large clouds of dust and pollen in the air
C wind moving the blades of a child's pinwheel
D wind moving a kite

41 What makes it possible for scientists around the world to compare scientific observations?

A using the same measuring system

B using both Celsius and Fahrenheit temperatures

C writing observations in different languages

D using sign language

42 Large piles of sand are found in the desert. The piles were formed mostly by

A water erosion

B wind erosion

C gravity

D gases from the air

43 Which of the following features would *not* help a plant in a dry environment?

A waxy coating on leaves to prevent water loss

B spiny leaves that do not allow water to evaporate

C spongy stems for holding water

D large leaves that can evaporate water quickly

44 The arrow on a wind vane points in the direction the wind is coming from. What direction is the arrow in the wind vane shown below pointing?

A north

B west

C south

D east

45 What are the three things that will protect you from spills during science class?

A sandals, raincoat, hat

B safety goggles, gloves, apron

C short sleeve shirt, jeans, leather shoes

D none of the above

46 How did sand get on the beaches near the oceans?

A A river left sand on all the beaches.

B The wind blew the sand to the beach.

C The ocean waves brought sand to shore and left it on the beach.

D Gravity forced the sand to flow to the beach from the nearby hills.

47 Which of the following *is* part of a frog's life cycle?

A hatchling

B tadpole

C pupa

D larva

48 A forest fire destroys all the trees in an area. After the fire, only grass and open fields are left. What changes in animal population might happen?

A Tree-dwelling animals like squirrels would increase.

B Grass-eating deer populations would increase.

C Duck populations would increase.

D Fish populations would increase.

49 You can find a wheel and axle in several places on a bicycle. Which bicycle part is not a part of a wheel and axle?

A handlebars
B back wheel
C pedal
D chain

50 A very old statue is in the park. Over the years, parts of it have crumbled and worn away. What is the likely cause of this happening?

A It was made of poor material.
B Sunlight destroyed it.
C Gravity made it wear away.
D Weathering by wind and rain wore parts of it away.

51 Six beads are shown below. How can these beads be classified into two groups with four beads in one group and two in another group?

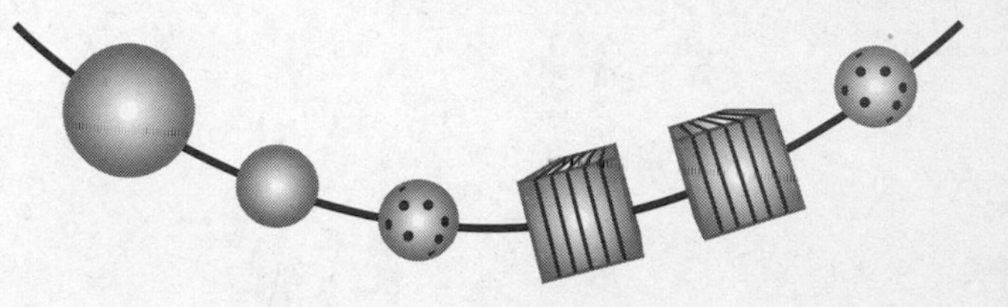

A large beads and small beads
B beads with stripes and beads with dots
C beads with dots and beads with no pattern
D round beads and square beads

52 What can you do at home to prepare for a science investigation?

A Skip breakfast and get to school early.
B Tie back long hair.
C Dress in loose clothing.
D all of the above

53 Which of the following is *not* a stage of growth and development for a butterfly?

A egg
B caterpillar
C pupa
D tadpole

54 A squirrel buries an acorn and it begins to grow into a tree. This activity is an example of

A animals depending on other animals for food
B plants depending on other plants for reproduction
C animals depending on humans
D plants depending on animals for reproduction

55 If you use two pulleys to lift a heavy load, instead of using only one pulley, the load will

A be harder to pick up
B be easier to pick up
C not change in how hard it is to pick up
D require the same force to be used as with an inclined plane

56 What important material is produced when rocks are weathered?

A water
B minerals
C soil
D magma

57 Which of the following should you do both before and after a science investigation?

A Write a report.
B Clean your work area.
C Read the investigation.
D Put on goggles.

58 Which of the following is a normal average life span of a human being?

A 11 months
B 2 years
C 80 years
D 75 months

---

59 Look at the drawing. Which two living things are more likely to compete for food?

A the bird and the man
B the two rabbits
C the dog and the rabbits
D the deer and the dog

---

60 How can a volcano have a positive effect on the environment?

A Erupting volcanoes can burn trees and homes.

B Erupting volcanoes can kill animals.

C Volcanic ash goes into the air.

D Volcanic ash turns into rich soil.

---

61 A student who lives in New York notices that every spring a lot of robins appear all of a sudden. What survival trait are the robins using?

A hibernation
B migration
C coloration
D camouflage

62 Which property of ice cream does *not* change when it is left at room temperature?

A shape
B state
C color
D hardness

63 Which of the following statements is true about the energy from the Sun?

A It is used by plants to make food.
B It causes water to evaporate.
C It causes winds on the Earth.
D all of the above

64 What force pulls you down a slide?

A friction
B magnetism
C gravity
D electricity

65 A student has been observing the growth of a tomato plant. The first week it is two inches tall. The second week it is three inches tall. The fourth week it is six inches tall.

The student could show this data best by using a

A graph
B chart
C drawing
D all of the above

66 The girl has trained the dog to sit before she feeds him. What kind of trait is this for the dog?

A an inherited trait
B a learned behavior
C a natural characteristic
D It is neither an inherited trait or a learned behavior.

67 At which stage is a caterpillar full grown?

A egg

B adult tadpole

C large larva

D adult moth

68 If trees grew taller over a few years, which of the following characteristics might give a giraffe an advantage in surviving?

A brown spots

B can run faster

C very long neck

D long eyelashes

69 Every year in the summer, strong rainstorms occur in Florida. What might happen in Florida if there were no summer storms?

A floods

B not enough water for crops and people

C the uprooting of trees

D the leveling of buildings

70 A doorbell will not stop ringing. Electricity is constantly being changed into sound energy. What could be wrong?

A The circuit is staying closed.

B The circuit is staying open.

C Electricity has stopped traveling in a circuit.

D Too much electricity is traveling in a circuit.

71 What is most likely happening when a tree loses all its leaves before a cold winter?

A It is dying.

B It is responding to a seasonal change.

C It is growing taller.

D It is beginning to grow fruit.

72 Some hurricanes cause much more damage than others do. Which hurricane in the chart below had the most costly damages?

HURRICANE NAME	DAMAGE COSTS
Hugo	$10 billion
Hazel	$281 million
Andrew	$20 billion
Audrey	$150 million
David	$1 billion

A Andrew
B Audrey
C Hazel
D Hugo

73 What does food provide for a growing animal?

A strength to go south for warmth
B energy for 3 hours
C energy and material to grow
D none of the above

74 Which of the following would *not* improve your science presentation?

A using a graph to show data
B using an illustration
C listening to suggestions from the class
D refusing to answer questions

75 You use a toaster to toast some bread. The toaster is changing electrical energy into

A chemical energy
B heat and light energy
C mechanical energy
D solar energy

76 Which of the following does *not* help an animal keep its body temperature normal when the seasons change?

A shedding fur
B increasing body fat
C releasing a scent
D growing fur

77 What do you think the fish might be sensing with their vision?

A danger
B a mate
C food
D warmer temperature

78 Staying in a building under the ground helps keep you safe during a

A sunny day
B flood
C tornado
D volcanic eruption

79 Which of the following do plants need to grow and survive?

A food
B space
C light
D all of the above

80 The angle at which the Sun's rays strike parts of Earth changes the

A daylight hours
B seasons
C time for a year
D phases of the Moon

81 A magnet will not attract a sheet of paper. So, why can you use a magnet to attach a sheet of paper to a refrigerator door?

A Magnetism can affect objects through a gas.
B Magnetism can affect objects through a liquid.
C Magnetism can affect objects through a solid.
D Magnetism works in outer space.

82 During which stage in the life cycle of a butterfly does metamorphosis occur?

A egg
B pupa
C larva
D adult

---

83 The picture below shows the phases of the Moon. Which statement about the phases of the Moon is true?

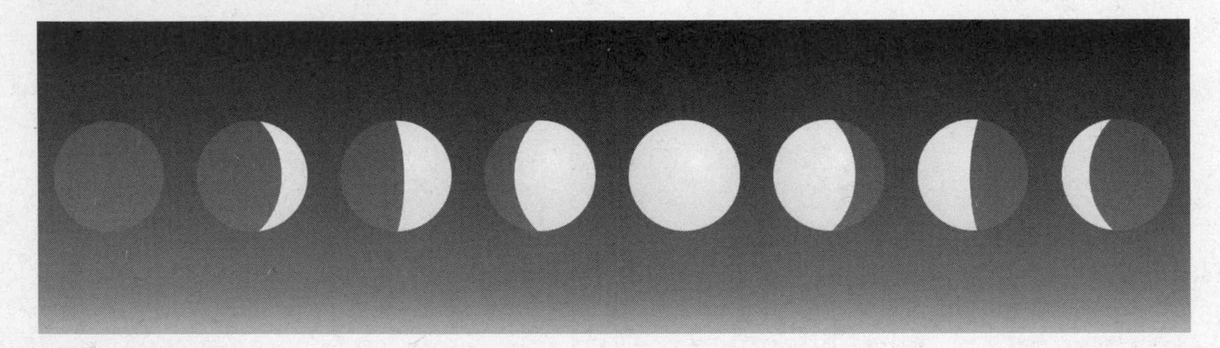

A The phases of the Moon are caused by Earth's shadow.
B The phases of the Moon are a pattern that occurs every month.
C The Moon always appears as a whole circle.
D When the Moon is in the new moon phase it is not in the sky at all.

---

84 Which of the following is both nonliving and part of the natural world?

A mountain
B plastic
C raccoon
D book

85 A student just completed presenting her science project. She grew one plant in sunlight and another in a closet. What would be a good question to ask?

A Did the plant in sunlight have better soil?

B Why did the plant in the closet turn yellow?

C Did the plant in sunlight need more water than the one in the closet?

D all of the above

86 Every year the Arctic fox's fur changes from white to brown. What environmental change causes this?

A Bears come out of hibernation.

B Winter becomes summer.

C Summer becomes winter.

D Autumn becomes winter.

87 Where might a plant store sugars and starches for future use?

A in its leaves

B in its roots

C in its fruit

D all of the above

88 Which of the following describes how living things are able to grow?

A Plants and animals convert energy from the Sun into food.

B Plants and animals use energy from water.

C Plants convert energy from the Sun into food, and animals get energy from plants that they eat.

D Plants get energy from animals, and animals convert energy from other animals.

89 When a wolf is far away from the rest of its pack, it howls. What behavior is this wolf showing?

A nest building

B hibernation

C migration

D communication

90 Which of the following ways is the correct way to find the average of these numbers?

5 14 9 4

A (14 + 5 + 9) divided by 4 = 7

B (9 + 5 + 14 + 4) divided by 4 = 8

C (9 + 5 + 14 + 4) multiplied by 4 = 128

D 14 + 9+ 5 + 4 = 32

91 In which way have humans changed the environment and hurt living things?

A built homes and roads where animals once lived

B hunted animals until some became extinct

C hiked in woods and left trash behind

D all of the above

92 The Sun appears to move across the sky because

A Earth revolves

B the Sun revolves

C Earth rotates

D the Sun rotates

93 Heat energy causes the particles in a solid to move faster and the object will get larger. Mr. Brown held a tight jar lid under hot running water. The heat from the water made the lid get larger than the jar. Then he was able to open the jar easily. This is an example of

A how a person uses matter that does not transfer heat very well
B how a person uses matter that transfers heat very well
C chemical energy interacting with mechanical energy
D mechanical energy interacting with the water

---

94 What is a possible result of smoking?

A high blood pressure
B lung damage
C increased risk of disease
D all of the above

---

95 Which of the following statements is true based on the chart?

	WEEK 1	WEEK 2	WEEK 3	WEEK 4
May 1999	3 days of rain	4 days of rain	3 days of rain	1 day of rain
May 2000	2 days of rain	3 days of rain	2 days of rain	2 days of rain
May 2001	2 days of rain	3 days of rain	2 days of rain	3 days of rain

A It has rained a total of 35 inches this summer.
B Every May, there are only five rainy days.
C For three years in a row, the second week of May was the rainiest.
D May is the rainiest month of the year.

Directions (96–153): For each question, write your answer in the space provided.

96 Why is it important not to use another person's towel?

____

____

97 How does the state of matter of wax in a candle change when the candle is lit?

____

____

98 Why are leaves important in the growth process of a plant?

____

99 A hill shows signs of erosion. Many paths have deepened from the top of the hill to the bottom. What caused the paths to get deeper?

____

____

100 Rick is growing a bean plant at school. For two days he is away from school on a field trip with his class. When he returns to school, he notices that the plant is wilted and the soil is dry. What generalization can Rick make based on his observations?

____

101 How does heat affect the liquid in a thermometer?

____

102 Name three ways to lead a life of good health.

1: ____________________

2: ____________________

3: ____________________

103 Name two behaviors that help animals survive cold months.

1: ____________________

2: ____________________

104 What two kinds of forces would a weight lifter use to lift barbells above his head?

1: ____________________

2: ____________________

105 Name two things a plant needs to produce food.

1: ____________________

2: ____________________

106 Name at least three characteristics that you were not born with.

1: ____________________

2: ____________________

3: ____________________

107 Explain a procedure that you could use to observe the weather near your home for one week.

____

____

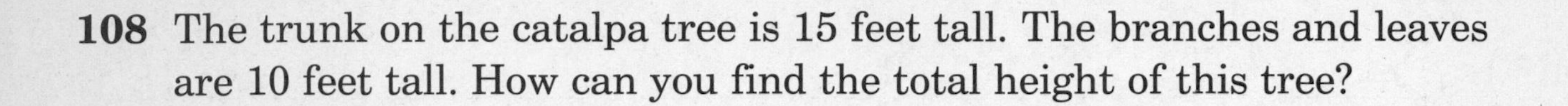

108 The trunk on the catalpa tree is 15 feet tall. The branches and leaves are 10 feet tall. How can you find the total height of this tree?

____

____

109 When we see a full moon, is the side facing away from the Earth dark or light? Draw a picture and explain your answer.

____

____

110 Use the greater than sign (>) to show the relationship between plant A and plant B. Use the less than sign (<) to show the relationship between plant B and plant C.

PLANT	HEIGHT
A	20 centimeters
B	10 centimeters
C	50 centimeters

Greater than: ______________________________

Less than: ______________________________

111 If two cars—car A and car B—appear to be the same size, how can you prove that car A weighs 50 pounds more than car B?

____

____

112 Alex is doing a supervised experiment in class. Alex places a thermometer in water and marks the time. Then, he heats the water until it is boiling, marking the time for every minute. Alex notices that the temperature of the water goes up while it is heating but once the water boils, the temperature remains the same. Explain two ways that this data can be displayed and which way would be better and why.

1: ____

____

2: ____

____

113 Sometimes cold foods are wrapped in newspapers to help keep them cold. Why are newspapers used for this purpose?

____

114 Sometimes the grass is very wet in the morning when there was no rain the night before. Explain where the water came from.

____

____

____

____

115 The Food Guide Pyramid contains the foods most needed in a balanced diet. Why is the candy bar, at the top, in the smallest part of the pyramid? Why are grains and breads in the largest part?

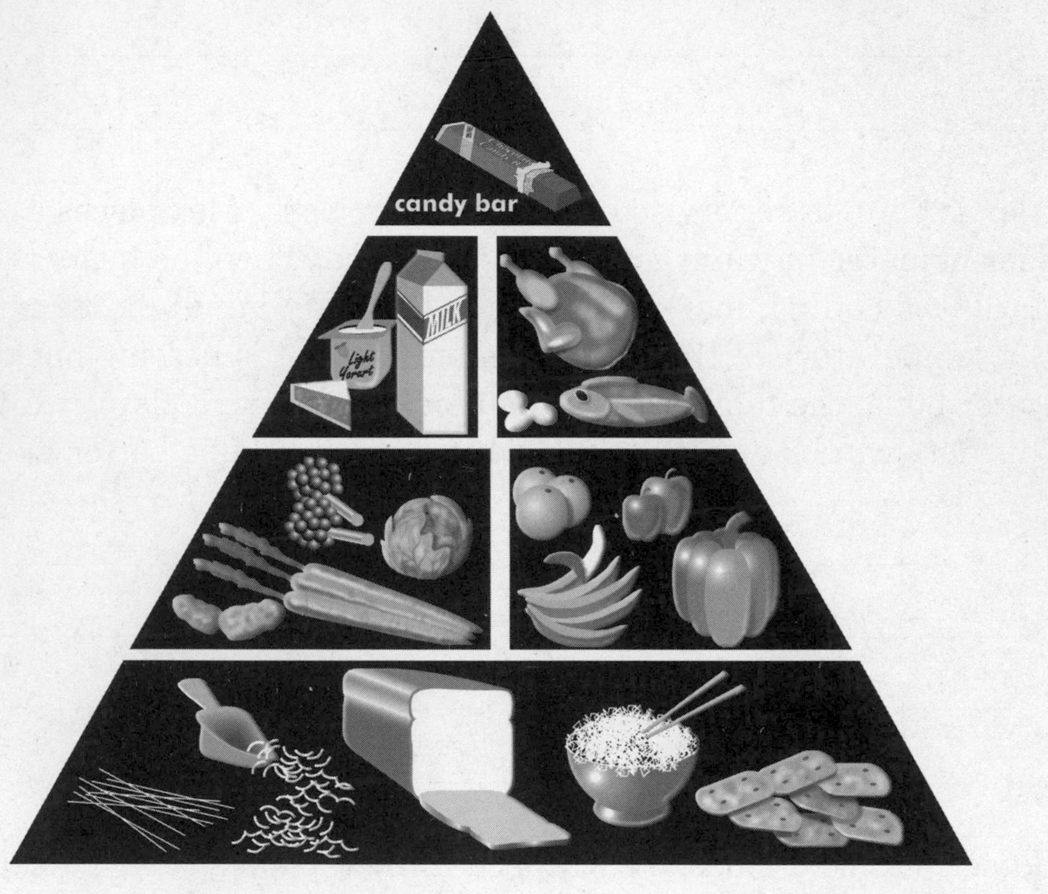

____

____

116 No rain has fallen for several months. How might this situation affect the groundwater in the area?

____

117 Ben is lifting boxes off the floor. He is putting them up on a shelf. How does the weight of the boxes affect how easily he can lift them?

____

118 There are two identical bowls of water, each with the same size piece of cork floating inside. The cork is floating higher in one bowl. What would be an appropriate question to ask in order to understand what is happening?

____

119 Name at least three stages of growth for a plant.

____

120 The front part of a boat, the bow, is a wedge. How does the bow help the boat use less force to move through water?

____

121 Explain what is needed to make a lightbulb light.

____

122 Why do you think polluted air would have an effect on a dog?

____

____

123 What might cause the killer whales in the Bering Sea to move to another body of water?

____

____

124 Cindy walks out of the hot sunlight into the cool shade of a tree. Why is the air in a shady spot cooler than the air in a sunny spot?

____

125 Suppose you added some pebbles to a full cup of water. How would the shape and volume of the water be affected?

____

____

126 Why do humans have the ability to live in almost any environment on Earth but animals cannot?

____

____

127 Why can't a covered container of gas be only half full?

____

128 How are magnetic and nonmagnetic objects sorted at a recycling center?

____

129 Mike spilled a box filled with small objects on the floor. Why should Mike try using a magnet to pick up the objects?

____

____

130 How does heat energy move? Give an example.

____

____

131 Which plant will most likely thrive? Why?

____

132 Electric energy travels in a closed circuit to light a bulb. How is the circuit affected if you open the switch?

____

133 How do you use force to draw? What force makes a pencil eraser work?

____

____

134 Amanda is making a leaf collection. What are three properties she could use to classify the leaves?

____

135 Describe the differences between the life span of the bean plant and the life span of the apple tree.

____

____

____

136 Describe the simple machines used below.

Yo-yo: ____________________

Swinging a bat: ____________________

137 A snake has a scaly skin that does not grow. Describe what probably happens to a snake so that it can grow.

____

____

138 Why are people told to lower their window shades on hot, sunny days?

____

139 Explain why electric cords are made of copper wire covered with rubber.

____

____

140 Scientists try to predict earthquakes. How can knowing when an earthquake will happen help save people's lives?

____

141 A basketball player needs to be able to stop suddenly on the court. Should the player wear shoes with rough soles or smooth soles?

____

142 What is the volume of the box shown below?

volume = length × width × height

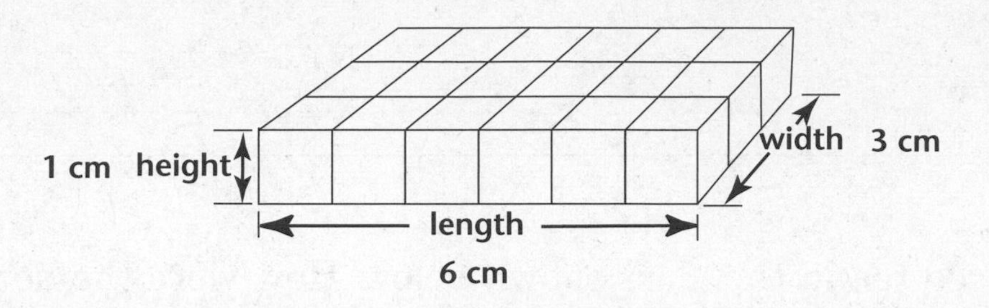

143 An electric appliance has a switch to turn it on and off. What does a switch do to a circuit?

144 Kim has 2 cubes of metal. They are the same shape. What properties can Kim check to determine what kind of metal each cube is made of?

145 It takes 10 minutes to drive to school from Ken's house. It takes 15 minutes to drive to the grocery store from his house. There are two traffic lights on the way to school and five on the way to the grocery store. Both locations are exactly 5 miles from Ken's house. How would you explain the difference in travel time?

146 One day on Earth is 24 hours long. Suppose the Earth spins around once every twelve hours instead of every 24 hours. How long would one day on Earth be?

____

147 There is iced tea in the glass shown below. How were the hot tea in the teapot and the ice cubes used to make the iced tea? How did heat move?

____

____

____

148 Insulators are materials that do not transfer energy well. Why are they put inside the outer walls of homes?

____

____

149 Carmen jumps as high as she can in the air. Why doesn't she stay up in the air?

____

150 How do you know that these two animals are the same species?

____

151 Name three ways that humans alter their environment.

____

____

____

152 What part of a flowering plant helps perform the life function of reproduction?

____

153 The boiling point of water is 100° Celsius. What could you use to find out if a pot of water was coming close to a boil?

____

Glossary

A

adapt the process by which species change over time in order to survive in the environment around them

average the typical number of a whole group of numbers

B

balanced diet a variety of healthy foods every day

batteries containers of matter with stored chemical energy

behavior the way an organism responds to a change in the environment

C

camouflage covering or color that helps animals look like their surroundings

circuit a path like a circle

classify to sort or arrange objects into groups based on how they are alike or how they are different

condensation the changing of water vapor (gas) into liquid water

D

data information gathered in an experiment

deposition the dropping or settling of eroded earth materials

E

energy what gives living things strength to live, grow, and carry out life processes; the ability to cause a change in matter

erosion the moving and carrying away of earth materials by wind and flowing water

erupt the occurrence whereby a volcano throws out gas, steam, ashes, dust, lava, or a combination of these

evaporation the changing of water (liquid) into water vapor (gas)

F

food chain a series of organisms through which energy is passed

force a push or a pull on an object

friction a rubbing force that resists the motion of one surface past another surface

G

germinates the process by which a seed starts to grow roots and shoots

graphic organizer a drawing, chart, or graph in which you can accurately display data

gravity a force of attraction between two objects

groundwater water that moves downward into the ground

growth the process by which plants and animals get bigger

H

harmful substances things such as drugs, alcohol, and tobacco that do damage to a body that will last a lifetime

hibernate to go to sleep for the winter, living on stored fat

I

inherited traits traits that are passed down from parents to offspring

L

lava hot liquid rock that comes out of a volcano

life cycle the stages of growth of a plant or an animal from the beginning of its life to the end of its life

life span the length of time from the beginning of a plant's or an animal's life until the end of its life

M

magnetic field the space around a magnet where the force of attraction is felt

magnetism force that may attract or repel certain materials

manufacture to make something

mass the amount of matter an object contains, measured in grams

math skills the ability to select the correct math operation to solve a problem

matter anything that takes up space and has mass, made up of particles that have properties that can be observed through our senses

mechanical energy the energy an object has because of its motion and the forces acting on it

metamorphosis the process of change in body size and shape of some animals as they develop

migrate to move from one place to another, usually with the change of seasons

N

nutrients substances a living thing needs for energy and growth

O

offspring new living things that parents produce, or the young of plants and animals

organism a living thing, such as a plant or an animal

P

perspire to release extra heat by letting water escape through the skin

pollution harmful substances that damage the air, water, land, or food supply

precaution something done to prevent an accident

precipitation water falling from the sky as rain, snow, sleet, or hail

predators animals that kill and eat other animals

prey animals hunted for food by other animals

producers organisms that use the Sun's energy to make their own food

property something that can be observed about an object, such as size, shape, color, hardness, taste, and mass

R

recording observations writing down the information gathered during an experiment

reproduction the process by which living things produce offspring

resource something living things need, such as food, air, or water

revolution Earth moving in a path around the Sun, resulting in one Earth year

rotation Earth spinning around once every twenty-four hours, resulting in day and night

runoff water flowing on Earth's surface

safety guidelines rules to follow when doing a science investigation

scientific inquiry the way scientists find facts and solve problems

scientist a person who has knowledge of science that is based on observed facts and tested truths

simple machines objects that make work easier by changing the strength, direction, or speed of a force

solar powered by energy from the Sun

species one of the groups into which scientists divide living things according to their shared characteristics

state of matter property that characterizes a substance as a solid, liquid, or gas

T

temperature a measure in degrees of how warm or cold a substance is

thrive to grow, do well, and be healthy

tools the instruments that help you conduct an experiment

traits qualities or characteristics of a living thing or a species

transfer to move from one person, thing, or place to another

V

vibrate to move back and forth rapidly

volcano an opening in the crust of Earth from which underground steam, ash, gas, and hot liquid rock escape

W

water cycle the constant movement of water from the ground, to the sky, and back again

water vapor water in its gas form

weather the condition of the outside air at a particular moment

weathering the breaking of rock into smaller rocks and soil

wind moving air

work the condition whereby a force makes something move through a transfer of energy

Scoring Rubric

Performance Levels

Level	Score Range	Description of Student Performance
4	85–100	**Meeting the Standards with Distinction** • Student demonstrates superior understanding of the intermediate-level science content and concepts for each of the learning standards and key ideas assessed. • Student demonstrates superior intermediate-level science skills related to each of the learning standards and key ideas assessed. • Student demonstrates superior understanding of the intermediate-level science content, concepts, and skills required for a secondary academic environment.
3	65–84	**Meeting the Standards** • Student demonstrates understanding of the intermediate-level science content and concepts for each of the learning standards and key ideas assessed. • Student demonstrates the science skills required for intermediate-level achievement in each of the learning standards and key ideas assessed. • Student demonstrates understanding of the intermediate-level science content, concepts, and skills required for a secondary academic environment.
2	44–64	**Not Fully Meeting the Standards** • Student demonstrates only minimal proficiency in intermediate-level science content and concepts in most of learning standards and key ideas assessed. • Student demonstrates only minimal proficiency in the skills required for intermediate-level achievement in most of the learning standards and key ideas assessed. • Student demonstrates marginal understanding of the science content, concepts, and skills required for a secondary academic environment.
1	0–43	**Not Meeting the Standards** • Student is *unable* to demonstrate understanding of the intermediate-level science content and concepts in most of the learning standards and key ideas assessed. • Student is *unable* to demonstrate the science skills required for intermediate-level achievement in most of the learning standards and key ideas assessed. • Student is *unable* to demonstrate evidence of the basic science knowledge and skills required for a secondary academic environment.

Science Reference Sheet

Equations

Acceleration (a)	=	$\frac{\text{change in velocity (m/s)}}{\text{time taken for this change (s)}}$	a	=	$\frac{v - v_o}{t}$
Average speed (v)	=	$\frac{\text{distance}}{\text{time}}$	v	=	$\frac{d}{t}$
Density (D)	=	$\frac{\text{mass (g)}}{\text{Volume (cm}^3\text{)}}$	D	=	$\frac{m}{V}$
Percent Efficiency (e)	=	$\frac{\text{Work out (J)}}{\text{Work in (J)}} \times 100$	%e	=	$\frac{W_{out}}{W_{in}} \times 100$
Force (F)	=	mass (kg) × acceleration (m/s^2)	F	=	ma
Frequency (f)	=	$\frac{\text{number of events (waves)}}{\text{time (s)}}$	f	=	$\frac{n \text{ of events}}{t}$
Momentum (p)	=	mass (kg) × velocity (m/s)	p	=	mv
Wavelength (λ)	=	$\frac{\text{velocity (m/s)}}{\text{frequency (Hz)}}$	λ	=	$\frac{v}{f}$
Work (W)	=	Force (N) × distance (m)	W	=	Fd

Units of Measure

m = meter	g = gram	s = second
cm = centimeter	kg = kilogram	Hz = hertz (waves per second)

J = joule (newton-meter)
N = newton (kilogram-meter per second squared)

Properties of Common Minerals

LUSTER	HARD-NESS	CLEAVAGE	FRACTURE	COMMON COLORS	DISTINGUISHING CHARACTERISTICS	USE(S)	MINERAL NAME	COMPOSITION*
Metallic Luster	1–2	✔		silver to gray	black streak, greasy feel	pencil lead, lubricants	**Graphite**	C
	2.5	✔		metallic silver	very dense (7.6 g/cm^3), gray-black streak	ore of lead	**Galena**	PbS
	5.5–6.5		✔	black to silver	attracted by magnet, black streak	ore of iron	**Magnetite**	Fe_3O_4
	6.5		✔	brassy yellow	green-black streak, cubic crystals	ore of sulfur	**Pyrite**	FeS_2
Either	1–6.5		✔	metallic silver or earthy red	red-brown streak	ore of iron	**Hematite**	Fe_2O_3
Nonmetallic Luster	1	✔		white to green	greasy feel	talcum powder, soapstone	**Talc**	$Mg_3Si_4O_{10}(OH)_2$
	2		✔	yellow to amber	easily melted, may smell	vulcanize rubber, sulfuric acid	**Sulfur**	S
	2	✔		white to pink or gray	easily scratched by fingernail	plaster of paris and drywall	**Gypsum** (Selenite)	$CaSO_4 \cdot 2H_2O$
	2–2.5	✔		colorless to yellow	flexible in thin sheets	electrical insulator	**Muscovite Mica**	$KAl_3Si_3O_{10}(OH)_2$
	2.5	✔		colorless to white	cubic cleavage, salty taste	food additive, melts ice	**Halite**	$NaCl$
	2.5–3	✔		black to dark brown	flexible in thin sheets	electrical insulator	**Biotite Mica**	$K(Mg,Fe)_3 AlSi_3O_{10}(OH)_2$
	3	✔		colorless or variable	bubbles with acid	cement, polarizing prisms	**Calcite**	$CaCO_3$
	3.5	✔		colorless or variable	bubbles with acid when powdered	source of magnesium	**Dolomite**	$CaMg(CO_3)_2$
	4	✔		colorless or variable	cleaves in 4 directions	hydrofluoric acid	**Fluorite**	CaF_2
	5–6	✔		black to dark green	cleaves in 2 directions at 90°	mineral collections	**Pyroxene** (commonly Augite)	$(Ca,Na)(Mg,Fe,Al)(Si,Al)_2O_6$
	5.5	✔		black to dark green	cleaves at 56° and 124°	mineral collections	**Amphiboles** (commonly Hornblende)	$CaNa(Mg,Fe)_4(Al,Fe,Ti)_3Si_6O_{22}(O,OH)_2$
	6	✔		white to pink	cleaves in 2 directions at 90°	ceramics and glass	**Potassium Feldspar** (Orthoclase)	$KAlSi_3O_8$
	6	✔		white to gray	cleaves in 2 directions, striations visible	ceramics and glass	**Plagioclase Feldspar** (Na-Ca Feldspar)	$(Na,Ca)AlSi_3O_8$
	6.5		✔	green to gray or brown	commonly light green and granular	furnace bricks and jewelry	**Olivine**	$(Fe,Mg)_2SiO_4$
	7		✔	colorless or variable	glassy luster, may form hexagonal crystals	glass, jewelry, and electronics	**Quartz**	SiO_2
	7		✔	dark red to green	glassy luster, often seen as red grains in NYS metamorphic rocks	jewelry and abrasives	**Garnet** (commonly Almandine)	$Fe_3Al_2Si_3O_{12}$

*Chemical Symbols: Al = aluminum, C = carbon, Ca = calcium, Cl = chlorine, F = fluorine, Fe = iron, H = hydrogen, K = potassium, Mg = magnesium, Na = sodium, O = oxygen, Pb = lead, S = sulfur, Si = silicon, Ti = titanium

✔ = dominant form of breakage

Commonly Used Units

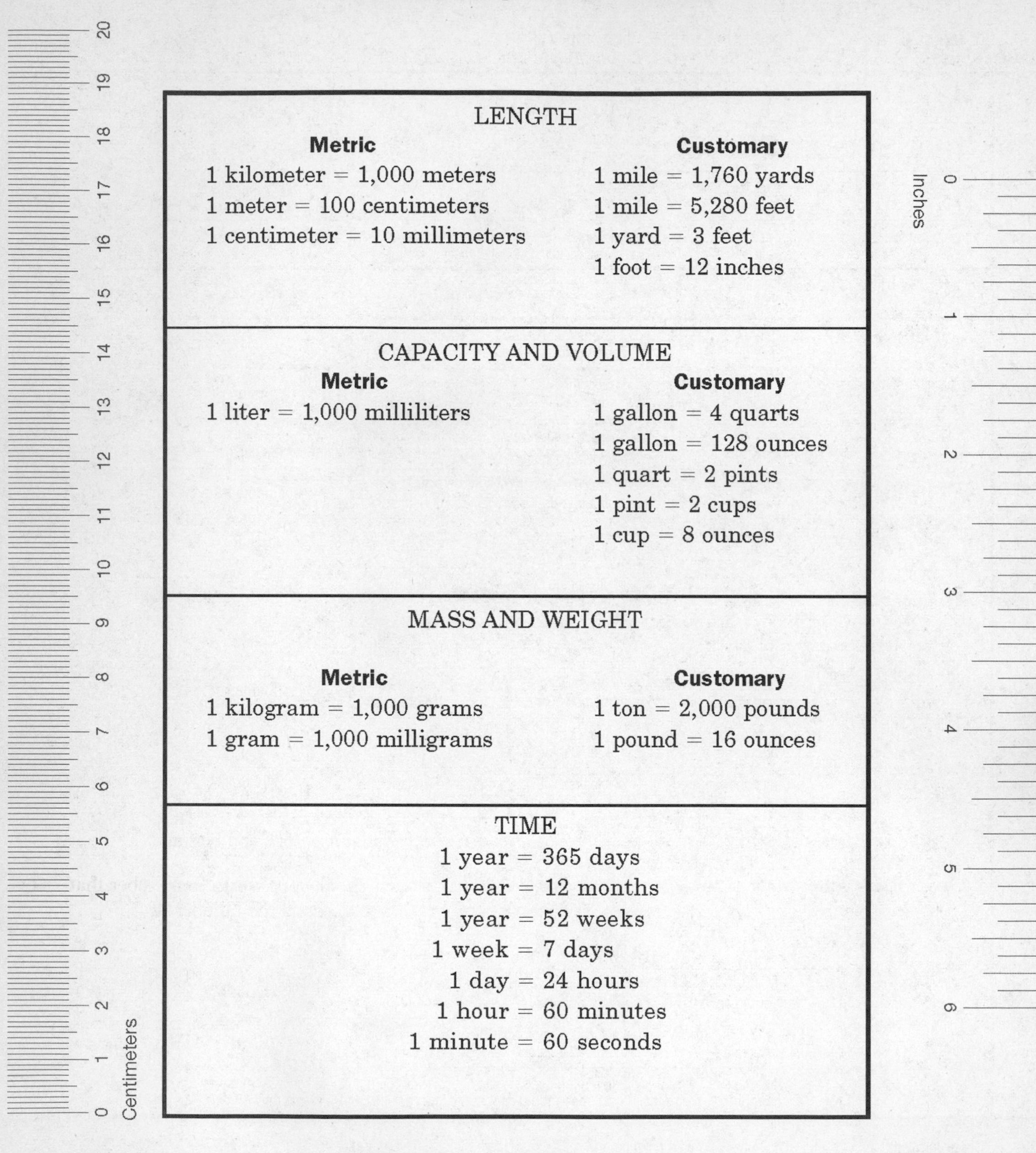

LENGTH

Metric	Customary
1 kilometer = 1,000 meters	1 mile = 1,760 yards
1 meter = 100 centimeters	1 mile = 5,280 feet
1 centimeter = 10 millimeters	1 yard = 3 feet
	1 foot = 12 inches

CAPACITY AND VOLUME

Metric	Customary
1 liter = 1,000 milliliters	1 gallon = 4 quarts
	1 gallon = 128 ounces
	1 quart = 2 pints
	1 pint = 2 cups
	1 cup = 8 ounces

MASS AND WEIGHT

Metric	Customary
1 kilogram = 1,000 grams	1 ton = 2,000 pounds
1 gram = 1,000 milligrams	1 pound = 16 ounces

TIME

1 year = 365 days
1 year = 12 months
1 year = 52 weeks
1 week = 7 days
1 day = 24 hours
1 hour = 60 minutes
1 minute = 60 seconds

Calculator Instructions

This is a picture of a generic calculator and its parts.

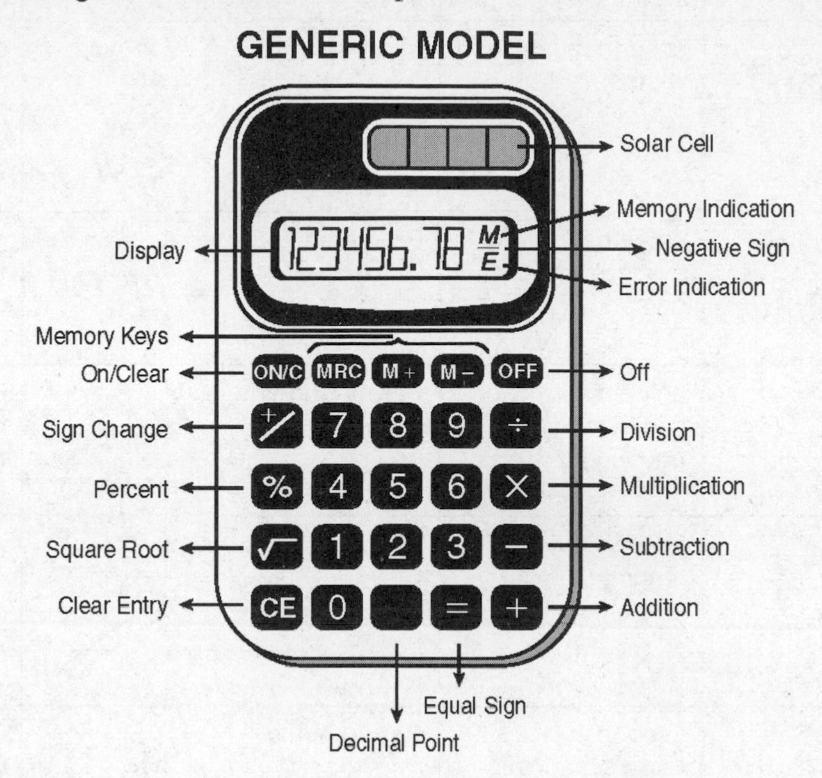

HELPFUL HINTS FOR TAKING THE SCIENCE TEST

1. Read the problem very carefully. Then decide whether or not you need the calculator to help you solve the problem.
2. When starting a new problem, always clear your calculator by pressing the clear key.
3. If you see an **E** in the display, clear the error before you begin.
4. If you see an **M** in the display, clear the memory and the calculator before you begin.
5. If the number in the display is not one of the answer choices, check your work. Remember that when computing with certain types of fractions, you may have to round the number in the display.
6. Remember, your calculator will NOT automatically perform the algebraic order of operations.
7. Calculators might display an incorrect answer if you press the keys too quickly. When working with calculators, use careful and deliberate keystrokes, and always remember to check your answer to make sure that it is reasonable.
8. The negative sign may appear either to the left or to the right of the number.
9. Always check your answer to make sure that you have completed all of the necessary steps.

Tools Commonly Used in Science

Tool	Description	Tool	Description
Anemometer	used to measure wind speed	**Microscope**	used to produce magnified images of small objects
Balance	used to measure the mass of objects	**Petri dish**	a container used to culture microorganisms
Beaker	an open cylindrical container with a pouring lip	**Safety goggles**	used in the laboratory to protect the eyes from hazardous materials
Calculator	used to do automatic math operations	**Scalpel**	an instrument used to dissect organisms
Compass	used to find Earth's magnetic north pole	**Spring scale**	used to measure weight and force
Computer	performs tasks by processing and storing information, including personal information	**Timer**	used to measure time
Graduated cylinder	a container marked in intervals that is used to measure volume and for finding density	**Thermometer**	used to measure temperature
Hot plate	an electrical plate that is used to heat objects	**Telescope**	used to produce magnified images of distant objects
Meterstick	a meter-long tool that is used to measure length	**Test tube**	a cylindrical tube, usually smaller than a beaker, that is open at one end and closed at the other; used to hold liquids